Advances in DUNE

Andreas Dedner, Bernd Flemisch,
and Robert Klöfkorn (Eds.)

Advances in DUNE

Proceedings of the DUNE User Meeting, Held in October 6th–8th 2010 in Stuttgart, Germany

Springer

Editors
Andreas Dedner
Mathematics Institute
University of Warwick
Coventry
United Kingdom

Robert Klöfkorn
Institut für Angewandte Analysis und
Numerische Simulation
Universität Stuttgart
Stuttgart
Germany

Bernd Flemisch
Institut für Wasser- und
Umweltsystemmodellierung
Universität Stuttgart
Stuttgart
Germany

ISBN 978-3-642-28588-2 e-ISBN 978-3-642-28589-9
DOI 10.1007/978-3-642-28589-9
Springer Heidelberg New York Dordrecht London

Library of Congress Control Number: 2012932487

Printed on acid-free paper

Springer is part of Springer Science+Business Media (www.springer.com)

Foreword by a DUNE Developer

Scientific Computing is a truly interdisciplinary endeavor where mathematicians, computer scientists and application scientists develop accurate, robust and scalable numerical simulators to solve challenging application problems. Albeit these groups have different tasks and aims in this process they would ideally share a common software framework. This should offer mathematicians the flexibility they need to implement ingenious new numerical schemes while computer scientists can concentrate on the performance aspects on modern hardware and application scientists can validate and test the new methods for real-world problems.

Nearly ten years ago in 2002, groups from Freiburg, Duisburg and Heidelberg, joined early on by a group from Berlin, started to develop the **D**istributed and **U**nified **N**umerics **E**nvironment (DUNE) with the ambition to provide such a software framework. Hereby, the attribute "distributed" meant both, distributed development and distributed (parallel) computing. From the start, the DUNE framework was based on three fundamental ideas:

1. Design general interfaces for the various concepts constituting the numerical solution process. In particular use interfaces to separate algorithms and data structures.
2. Use C++ template programming techniques to mitigate the performance burden imposed by a general, fine-grained interface.
3. Provide implementations based on legacy codes in order to allow the reuse of existing tested and well-known codes.

These ideas proved to be very fruitful until the present day. A mature state of the core modules covering computational grids as well as sparse linear algebra and iterative solvers was reached in 2007 with the release of version 1.0. Still, the flexibility with respect to different types of computational grids is one of the biggest assets of using DUNE. In recent years much work has gone into interfaces for function spaces and discretization schemes. Moreover, DUNE is increasingly used by other groups and individuals as the basis of their work.

This development is very well reflected by the collection of articles contained in this volume which have been presented at the first DUNE user meeting. They cover extensions of the software framework, in particular several "meta grids", the implementation of particular numerical schemes as well as the integration of DUNE into other larger software packages. It is particularly encouraging that the initiative for organizing the first user meeting was taken by the users themselves. Clearly, an active user community contributing through feedback, testing, bug fixes and organization of events is the criterion for a successful software project.

I would like to thank the organizers of the first DUNE user meeting for their efforts and sincerely hope that there will be more such exciting and stimulating events in the future!

Heidelberg, September 2011 Peter Bastian

Foreword by a DUNE User

If the Distributed Unified Numerics Environment, aka. DUNE, did not exist, it would have to be invented. While partial differential equations (PDE) have been studied with analytical techniques for centuries, their analysis was restricted to comparatively simple geometries, material systems and boundary conditions. Only the relatively recent advent of powerful computing infrastructure, such as fast processors with multiple cores, large amounts of memory, huge hard disk capacity for permanent storage, efficient communication networks and highly impressive visualization hardware has enabled the numerical solution of PDEs for complicated geometrical domains, realistic material systems and boundary conditions, at least in principle.

Of course, over the last 30 years or so, a wealth of numerical codes has emerged, addressing a wide span of physical fields from fluid dynamics, acoustics, thermodynamics, mechanics, soil physics, quantum mechanics and computational electromagnetics, to name just a few. The numerical solution of such problems usually reduces to the spatial discretization of the differential operators on some uniform (regular Cartesian or hexahdedral) or on unstructured (tetrahedral, pyramidal, prismatic) meshes. Common to both approaches is the idea of formulating the original PDE in a simplified way. For example, in the case of a regular Cartesian grid the PDE is approximated via the finite difference formulation of the original differential operator; in the case of structured or unstructured meshes, the solution of the original PDE is written as a superposition of known functions, aka. basis functions, scaled by yet unknown coefficients, aka. degrees of freedom (DoF) for every finite element (FE) of the mesh. Depending on the type of the PDE and its discretization linear systems of equations described by matrices with variable structural properties eventually need to be solved. While all these steps appear relatively simple on paper, their actual realization in the form of a working computer program represents a formidable task of considerable complexity. Starting with the generation of a suitable mesh, its handling in the computer's memory, or even more challenging, its partitioning on a parallel computer system with distributed memory architecture, the formulation of basis functions for all the topological components of a finite element mesh (nodes, edges, faces, the element itself), the subsequent solution of a very large, sparse linear equation system and, eventually the visualization of the numerical solution in time and one-, two- or three-dimensional space, the implementor of a numerical computer code is confronted, over and over again, with the same or similar series of elementary tasks. While the actual realization of these tasks is cumbersome and sometimes error prone, it nevertheless requires care and patience. A programming flaw during these steps easily keeps an implementor busy for weeks when debugging a complex finite element code. On the other hand, in today's highly competitive scientific world where publication of novel, unheard of methods is the clear priority, such basis work is simply not publishable but must be accounted for essentially as an initial investment, necessary for the ultimate prize, namely the study of new methods for promising new fields of applications.

All the more laudable is therefore the DUNE development effort since, almost exclusively, DUNE provides an almost perfect basis for the rapid realization of a new method in a wide span of physical fields, their common link being the loosely defined notion of finite difference of finite element methods. The DUNE project is a great achievement and a generous offer to the community concerned with research into the next generation of numerical methods for the analysis of physics governed by PDEs. It is indeed an honor and a great pleasure for me to write the foreword of the 2010 DUNE User Group Meeting. The span of physical problems presented in the report speaks for itself. I myself, have overseen the development of a computational electromagnetics solver family, HADES, based on DUNE, exploiting in particular its finite element infrastructure. While I am fully aware that the DUNE effort now rests on many strong and able shoulders, I wish to mention that in the beginning there were only a few people working on the concept and supporting its continued and sustained development. While I certainly appreciate the work done by the developers of DUNE and associated libraries, such as grid management etc., and also, in the sense of quality assurance, the users, it is my deep conviction and always has been that great achievements rarely result from committee work but from great minds. Thus, let us be thankful for the DUNE effort, use it for ever more sophisticated numerical methods and hope for many happy returns of DUNE user group meetings.

Crete, September 2011 Benedikt Oswald

Preface

This volume contains 13 refereed papers presented at the First DUNE User Meeting held at Stuttgart, Germany, October 6-8, 2010. The meeting brought together almost 30 DUNE users and developers from Germany, Austria, U.K., and Norway. The following talks were presented during the meeting:

- Andreas Dedner (University of Warwick, U.K.): Shape functions based on the generic reference elements,
- Martin Drohmann (University of Münster, Germany): Computing reduced basis solutions with an implementation interface connecting DUNE with Matlab,
- Bernd Flemisch (University of Stuttgart, Germany): Upscaling of porous media flow with DuMu^x and DUNE-MULTIDOMAINGRID,
- Christoph Gersbacher (University of Freiburg, Germany): The DUNE-PRISMGRID module,
- Wolfgang Giese (Humboldt University Berlin, Germany): Use of DUNE on a Linux-Cluster with a biological application,
- Sebastian Götschel (Zuse Institute Berlin, Germany): State trajectory compression for optimal control with parabolic PDEs,
- Christoph Grüninger (University of Stuttgart, Germany): Using DUNE and PDE-LAB for the diploma thesis as a student,
- Patrick Henning (University of Münster, Germany): The heterogeneous multi-scale finite element method and its implementation using DUNE,
- Robert Klöfkorn (University of Freiburg, Germany): Higher order adaptive and parallel simulations with DUNE,
- Andreas Lauser (University of Stuttgart, Germany): The DuMu^x material law framework,
- Chamakuri Nagaiah (University of Graz, Austria): Optimal control of the reaction-diffusion equations in electrophysiology,
- Steffen Müthing (University of Stuttgart, Germany): Solving multi-domain problems with PDELAB and DUNE-MULTIDOMAIN,
- Martin Nolte (University of Freiburg, Germany): Performance pitfalls to be avoided in the DUNE grid interface,

- Atgeirr Rasmussen (SINTEF Oslo, Norway): Usage of DUNE at SINTEF Applied Mathematics,
- Maik Schenke (University of Stuttgart, Germany): On the Analysis of Porous Media Dynamics using a DUNE-PANDAS Interface,
- Christian Waluga (RWTH Aachen, Germany): Hybridized DG methods: theory and implementation in DUNE,
- Jonathan Youett (Freie Universität Berlin, Germany): An implementation of biomechanics via DUNE.

Most of these talks have evolved into contributions to these proceedings. These contributions are grouped into three parts: "Core DUNE Extensions," "External DUNE modules," and "Specific Methods and Applications."

We would like to thank all the authors for their contributions to this proceedings and all the referees for ensuring a high quality of the manuscripts. Furthermore, we gratefully acknowledge the financial support of the First DUNE User Meeting provided by the Cluster of Excellence "Simulation Technology" located at the University of Stuttgart. Finally, we would like to thank Thomas Ditzinger and Holger Schäpe at the Springer-Verlag for their competent and kind guidance through the book project.

Stuttgart and Warwick,
October 2011

Andreas Dedner
Bernd Flemisch
Robert Klöfkorn

Contents

Part I
Core DUNE Extensions

Construction of Local Finite Element Spaces Using the Generic Reference Elements

Andreas Dedner and Martin Nolte

Abstract. Based on the recursive definition of the generic reference elements in DUNE-GRID, we present an effective framework for the implementation of finite element shape functions. Such a shape function set is described by a set of functionals defining the degrees of freedom and an arbitrary basis of the finite element space on the reference element. To illustrate the power of this approach we show how Lagrange shape functions, Raviart-Thomas shape functions, and L^2-orthonormal shape functions fit into this framework.

1 Introduction

The implementation of finite element functions is an important, yet tedious task in the construction of finite element methods. Especially high order spaces with many degrees of freedom require a high implementational effort if a numerically stable implementation is to be provided. This holds especially for class C^k spaces with $k > 0$ or vector valued elements such as the Raviart-Thomas element or the Nedelec element. Many finite element packages thus only provide low order finite element spaces; high order spaces are at best provided for simple Lagrange spaces or for tensor function spaces where the high order only needs to be implemented in 1D (e.g., www.dealii.org, www.freefem.org, home.gna.org/getfem, hpfem.org/hermes).

In this paper we present an effective method for the implementation of high order shape functions. Similar to the introduction of local finite elements in [5, 4], our implementation only requires the description of the function space and a set of

Andreas Dedner
Mathematics Institute, University of Warwick, Coventry CV4 7AL, UK
e-mail: `A.S.Dedner@warwick.ac.uk`

Martin Nolte
Abteilung für Angewandte Mathematik, Universität Freiburg, Hermann-Herder-Str. 10, 79104 Freiburg i. Br., Germany
e-mail: `nolte@mathematik.uni-freiburg.de`

functionals describing the degrees of freedom (DoF) on some reference element R. With an additional assignment of the DoFs to subentities of the reference element, a local finite element in the sense of DUNE-LOCALFUNCTIONS [10] can be constructed automatically.

In [6] a general description of grid based discretization schemes is provided. Especially *localized discrete function spaces* (see Definition 7 in [6]) play a crucial role in many efficient numerical methods. These are build on top of special basis function sets, so-called shape function sets , which are defined on the reference elements as they are provided by DUNE-GRID. The local finite element classes in DUNE-LOCALFUNCTIONS provide for a given reference element R a tuple $(\hat{\mathscr{B}}_R, \boldsymbol{\Lambda}_R, \hat{\mathscr{C}}_R)$. Here $\boldsymbol{\Lambda}_R = (\lambda_j)_j$ are the *nodal variables* and $\hat{\mathscr{B}}_R = (\hat{\phi}_i)_i$ is a set of basis functions $\hat{\phi}_i : R \to \mathbb{R}^r$ satisfying $\lambda_j(\hat{\phi}_i) = \delta_{ij}$. The $\hat{\mathscr{C}}_R$ assign to each functional a subentity of the reference element and are used in the construction of global finite element spaces V_h on a grid. The nodal variables play a crucial role in basically all properties of the local finite element space. In fact, [5] defines affine families of finite elements only through the nodal variables and the finite dimensional function space $\hat{V}_R = \mathrm{span}(\hat{\mathscr{B}}_R)$. Given a set of nodal variables it is often quite complicated to construct the specific basis set $\hat{\mathscr{B}}_R$, while it can be quite simple to provide any basis set for V_R. Assume now that we are given a basis $\hat{\boldsymbol{B}}_R = (\hat{\psi}_0, \dots, \hat{\psi}_{N_R-1})$ of V_R and a set of nodal variables $\boldsymbol{\Lambda}_R = (\lambda_0, \dots, \lambda_{N_R-1})$ satisfying

- $\mathrm{span}(\hat{\boldsymbol{B}}_R) = \hat{V}_R$.
- λ_j is a linear functional on $\hat{V}_R$.
- the matrix $A_R = \big(\lambda_j(\hat{\psi}_i)\big)_{ij} \in \mathbb{R}^{N_R \times N_R}$ is regular.

Then the basis $\hat{\mathscr{B}}_R = \{\hat{\phi}_0, \dots, \hat{\phi}_{N_R-1}\}$ satisfying $\lambda_j(\hat{\phi}_i) = \delta_{ij}$ is given by

$$\hat{\mathscr{B}}_R = (A_R)^{-T} \hat{\boldsymbol{B}}_R. \tag{1}$$

A second type of basis functions, which does not directly fit into the concept described above, is defined through a symmetric, positive definite bilinear form $a_R(\cdot,\cdot)$ instead of a set of functionals $\boldsymbol{\Lambda}_R$. We now seek a basis $\hat{\mathscr{B}}_R = (\hat{\phi}_0, \dots, \hat{\phi}_{N_R-1})$ spanning a discrete function space V_R which is orthonormal with respect to a_R, i.e., $a_R(\hat{\phi}_i, \hat{\phi}_j) = \delta_{ij}$. It is easy to see that given a basis $\hat{\boldsymbol{B}}_R = (\hat{\psi}_0, \dots, \hat{\psi}_{N_R-1})$ of V_R, the required basis is again of the form (1). The matrix A_R is given by

$$A_R A_R^T = M_R := \big(a_R(\hat{\psi}_j, \hat{\psi}_i)\big)_{ij}, \tag{2}$$

i.e., A_R is the *square root* of the symmetric and positive definite matrix M_R. It can, for example, be constructed by a Gram-Schmidt procedure. The corresponding set of nodal variables $\boldsymbol{\Lambda}_R = (\lambda_0, \dots, \lambda_{N_R-1})$ is then constructed a posteriori from the basis $\hat{\mathscr{B}}_R$ by setting $\lambda_i(f) := a(f, \hat{\phi}_i)$.

In the following section we describe the steps required to construct a wide range of shape function sets using some utility classes found in the module DUNE-LOCALFUNCTIONS. We will concentrate on the construction of the set of reference elements $\hat{\mathscr{R}}^d$, the *pre basis* $\hat{\boldsymbol{B}}_R$, and detail how the pre basis together with the nodal variables (or the bilinear form) can be used to construct $\hat{\mathscr{B}}_R$. Examples for this type of shape function sets are given in Section 3. Some numerical experiments are shown at the end of the paper.

2 Construction of Local Finite Elements

The next three subsections describe the construction and give some implementational details for obtaining the set of reference elements $\hat{\mathscr{R}}^d$, the *pre basis* $\hat{\boldsymbol{B}}_R$, and the *basis transformation matrix* A_R^{-T}.

In DUNE-LOCALFUNCTIONS the basis $\hat{\mathscr{B}}_R$ is modeled by the class `LocalBasis` while the functionals $\boldsymbol{\Lambda}_R$ are represented by the class `LocalInterpolation`. The pre basis $\hat{\boldsymbol{B}}_R$ is an auxiliary class.

To describe our generic construction process for shape functions, we first require some notation from the DUNE-GRID interface. For general definitions, we refer to [2].

Let $\mathscr{T}_h$ be a tessellation of $\Omega \subset \mathbb{R}^w$. For each element $T \in \mathscr{T}_h$ we assume that there exists a reference element $R_T \subset \mathbb{R}^d$ in a small set $\hat{\mathscr{R}}^d$ of reference elements and a bijective reference mapping $F_T : R_T \to T$ which is assumed to be at least of class C^1.

The subentities of codimension $c = 0,\dots,d$ in $R \in \hat{\mathscr{R}}^d$ are the elements of the set $\hat{E}_R^c$. We assume that for each $\hat{e} \in \hat{E}_R^c$ there is a reference element $R_{\hat{e}} \in \hat{\mathscr{R}}^{d-c}$ and a reference embedding $i_{\hat{e}} : R_{\hat{e}} \to \hat{e}$. The subentities of codimension c in $T \in \mathscr{T}_h$ are given by $F_T(\hat{e})$ with $\hat{e} \in \hat{E}_{R_T}^c$. We denote the set of all subentities of codimension c of $T \in \mathscr{T}_h$ by E_T^c. For a subentity $\hat{e} \in \hat{E}_R^1$ we further denote by $n_{\hat{e}}$ the outer normal to R on $\hat{e}$.

In DUNE, the tessellation $\mathscr{T}_h$ is modeled by a `GridView`. The set of all reference elements $\hat{\mathscr{R}}^d$ is represented by the `GenericReferenceElementContainer` and the reference embedding $i_{\hat{e}}$ is given by the method `mapping<codim>` on the reference element. The geometric mapping F_T is described by the class `Geometry`.

2.1 *The Generic Reference Elements*

The set of reference elements $\hat{\mathscr{R}}^d$ is constructed based on a recursion over the space dimension d.

Definition 1. Given a set of reference elements $\hat{\mathscr{R}}^d$ with $R \subset \mathbb{R}^d, R \in \hat{\mathscr{R}}^d$ we define $\hat{\mathscr{R}}^{d+1} = \{R^|, R^\circ : R \in \hat{\mathscr{R}}^d\}$, where for $R \in \hat{\mathscr{R}}^d$:

$$R^| = \{(x,z) : x \in R, z \in [0,1]\}, \qquad R^\circ = \{(x(1-z),z) : x \in R, z \in [0,1]\}.$$

For $d = 0$ we set $\hat{\mathscr{R}}^0 = \{P\}$ with $P = \{0\} \subset \mathbb{R}^0$.

Note that R° is the Duffy transform of $R^{|}$. For the 1D reference elements there is a non-uniqueness issue, since $P^{|} = P^{\circ}$; this carries over to all higher dimensions, i.e., $P^{||} = P^{\circ|}$, $P^{\circ\circ} = P^{|\circ}$, etc. The recursive construction of reference elements is sketched in Fig. 1.

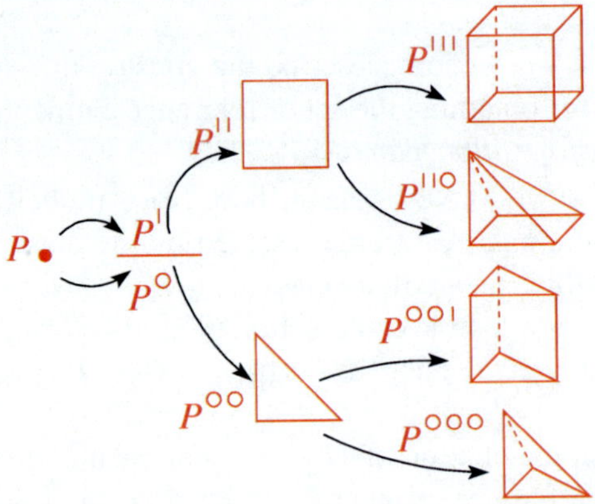

Fig. 1 Recursive Construction of the Generic Reference Elements up to 3d.

In the code, this recursive definition is mapped into a template recursion. The three classes

```
struct Point;
template <class Base> struct Prism;
template <class Base> struct Pyramid;
```

represent P, $R^{|}$, and R°, respectively. They are referred to as the *topology* of the reference element. In addition, each reference element has a dynamic identifier, its `topologyId`. This is an integer defined through:

$$\begin{aligned} \texttt{topologyId}(P) &= 0, \\ \texttt{topologyId}(R^{|}) &= \texttt{topologyId}(R) + 2^{d-1}, \\ \texttt{topologyId}(R^{\circ}) &= \texttt{topologyId}(R). \end{aligned}$$

A reference element $R \in \hat{\mathscr{R}}^d$ is thus uniquely determined by its `topologyId` and dimension. Note that, due to the non-uniqueness issue described above, each reference element is given by two topology ids differing in the lowest digit of their binary representation. Moreover, the valid topology ids are exactly the integers $0,\ldots,2^d-1$.

The recursive definition of the reference elements is also used to provide the numbering of the subentities. Furthermore, it induces quite naturally the reference embeddings $i_{\hat{e}}$ for $\hat{e} \in \hat{E}^c_R$.

There are two special classes of reference elements, *cubes* which are of the form $R = P^{|\cdots|}$ and *simplices* of the form $R = P^{\circ\cdots\circ}$. The topology id of a simplex is always equal to zero (or one) while a d-dimensional cube has the topology id $2^d - 1$ (or $2^d - 2$).

2.2 The Pre Basis

In the following we assume that the pre basis $\hat{\boldsymbol{B}}_R = (\hat{\psi}_0, \dots, \hat{\psi}_{N_R-1})$ is given by

$$\hat{\psi}_i(x) = \sum_{l=0}^{n_R-1} b_{i,l} m_l(x)$$

for some set $M_R = (m_0, \dots, m_{n_R-1})$ of scalar functions and coefficients $b_{i,l} \in \mathbb{R}^r$.

In DUNE-LOCALFUNCTIONS we have so far implemented for each reference element $R \in \hat{\mathscr{R}}^d$ and each order k one set of "monomials" $\mathscr{M}_R^k = \bigcup_{q \le k} \bar{\mathscr{M}}_R^q$. These are aimed at an easy implementation of Lagrange type basis functions, leading to B_R being the identity matrix. The definition (and implementation) of these sets follow the recursive construction of the reference elements $\hat{\mathscr{R}}^d$. In the special case of a simplex type reference elements R, the set $\bar{\mathscr{M}}_{R^\circ}^q$ contains exactly the monomials x^α with $|\alpha| = q$ while in the case that R is a cube we have that functions in $\bar{\mathscr{M}}_{R|}^q$ are of the form x^α with $\max_i \alpha_i = q$.

Definition 2. For $d = 0$ we have $\bar{\mathscr{M}}_P^0 = \{1\}$ and $\bar{\mathscr{M}}_P^q = \emptyset$ for $q > 0$. Then the basis functions in $\bar{\mathscr{M}}_{R^\circ}^q$ for $R \in \hat{\mathscr{R}}^d$ are given by

$$\bar{\mathscr{M}}_{R^\circ}^q = \{(1-z)^{q-i} z^i \psi\big(\tfrac{1}{1-z} y\big) : \psi \in \bar{\mathscr{M}}_R^{q-i},\ i = 0, \dots, q\}$$

with $y \in R$ and $(y,z) \in R^\circ$. For $y \in R$ and $(y,z) \in R^|$ we define the sets

$$\bar{\mathscr{M}}_{R|}^q = \{z^{q+i}\psi(y) \colon \psi \in \bar{\mathscr{M}}_R^q,\ i = -q, \dots, 0\} \bigcup \{z^q \psi(y) \colon \psi \in \bar{\mathscr{M}}_R^{q-i},\ i = 1, \dots, q\}.$$

for $q = 0, \dots, k$.

The recursion leads to a natural ordering of the functions in $\mathscr{M}_R^k$ so that we obtain a vector $(m_{R,i}^k(x))_i$ for each reference element $R \in \hat{\mathscr{R}}^d$ and $k \ge 0$.

Note that in the case of a general pyramid, like the 3-dimensional pyramid $\mathrm{P}^{||\circ}$, these "monomials" are actually rational functions rather than polynomials.

The recursion relations can be summarized in the following tables

	$i=0$	$i=1$	$i=2$
$\bar{\mathscr{M}}_{R^\circ}^0$	$\bar{\mathscr{M}}_R^0$		
$\bar{\mathscr{M}}_{R^\circ}^1$	$(1-z)\bar{\mathscr{M}}_R^1$	$z\bar{\mathscr{M}}_R^0$	
$\bar{\mathscr{M}}_{R^\circ}^2$	$(1-z)^2\bar{\mathscr{M}}_R^2$	$(1-z)z\bar{\mathscr{M}}_R^1$	$z^2\bar{\mathscr{M}}_R^0$

	$i=-2$	$i=-1$	$i=0$	$i=1$	$i=2$
$\bar{\mathscr{M}}_{R\|}^0$			$\bar{\mathscr{M}}_R^0$		
$\bar{\mathscr{M}}_{R\|}^1$		$\bar{\mathscr{M}}_R^1$	$\bar{\mathscr{M}}_R^1 z$	$\bar{\mathscr{M}}_R^0 z$	
$\bar{\mathscr{M}}_{R\|}^2$	$\bar{\mathscr{M}}_R^2$	$\bar{\mathscr{M}}_R^2 z$	$\bar{\mathscr{M}}_R^2 z^2$	$\bar{\mathscr{M}}_R^1 z^2$	$\bar{\mathscr{M}}_R^0 z^2$

The implementation is based on the template class

```
template< class Topology, class Field > class MonomialBasis;
```

which provides methods for evaluating the basis $\mathscr{M}_R^k$ and its derivatives. A second template class

```
template< int d, class Field > class VirtualMonomialBasis;
```

provides a virtual interface for evaluating $\mathscr{M}_R^k$ for any $R \in \hat{\mathscr{R}}^{\mathrm{d}}$ where the topology id of R is passed to the constructor.

2.3 The Local Finite Element Implementation

The construction of a `LocalFiniteElement` in DUNE-LOCALFUNCTIONS requires the implementation of the `LocalBasis`, the `LocalInterpolation`, and the `LocalCoefficients` [10].

We have seen that, given an implementation of some function set $\mathscr{M}_R^k = \{m_i\}_{i=0}^{n_{R,k}-1}$, the local basis can be evaluated by applying the matrix operation

$$\hat{\mathscr{B}}_R = (A_R)^{-T} B_R \mathscr{M}_R^k,$$

where the pre basis satisfies $\hat{\boldsymbol{B}}_R = B_R \mathscr{M}_R^k$ with $B_R = (b_{ilj})_{ijl} \in \mathbb{R}^{N_R \times r \times n_{R,k}}$.

Provided that one of the available implementations of $\mathscr{M}_R^k$ described in the previous section suffices, only the matrix B_R and the nodal variables need to be provided. The main work in the implementation of a new shape function set is thus the construction of the `LocalInterpolation`. In addition to the interface method

```
template < class Function >
void interpolate(const Function& f,std::vector<Field>& out) const
```

we need an additional method

```
template < class Basis, class Matrix >
void interpolate(const Basis& f,Matrix& out) const
```

used to fill the matrix A from the given set of basis functions. There is a helper class `InterpolationHelper` which allows to use a single implementation for both versions of the method `interpolate`.

Setting up the basis thus requires evaluating the nodal variables $\boldsymbol{\Lambda}_R$ for $\hat{\boldsymbol{B}}_R = B_R \mathscr{M}_R^k$ and inverting A_R. Evaluating the basis $\hat{\mathscr{B}}_R$ in some point requires the evaluation of $\mathscr{M}_R^k$ and multiplying the result by $(A_R)^{-T} B_R$. Notice that the construction of $(A_R)^{-T} B_R$ needs to be performed only once during a simulation. Furthermore, our implementation of numerical schemes caches the evaluation of $\hat{\mathscr{B}}_R$ in quadrature points (see [6]). We thus assume that both, construction and evaluation of the basis, are not runtime critical tasks. Especially during the construction phase stability with respect to round-off errors is essential, since we have to invert A_R, e.g., by a Gauss-Jordan algorithm. Thus, we allow the use of high precision floating point classes as, for example, provided by the GMP [7] or ALGLIB [3]. Each local finite element class has two template arguments. The type `ComputingField` specifies the floating point type used during the construction phase, while the second type, `StorageField`, specifies the floating point type used for storing $(A_R)^{-T} B_R$, for evaluating $\mathscr{M}_R^k$, and for the final multiplication $(A_R)^{-T} B_R \mathscr{M}_R^k$.

The whole construction process of local finite element spaces given the local interpolations $\boldsymbol{\Lambda}_R$ and the pre bases $\hat{\boldsymbol{B}}_R$ is carried out by the class

```
template< class PreBasisFactory, class InterpolationFactory,
          unsigned int dim, unsigned int dimR,
          class StorageField, class ComputingField >
struct DefaultBasisFactory;
```

The first two template arguments are factory classes providing the pre basis and the nodal variables for a given reference element. They should be derived from the `TopologyFactory` helper class implemented in DUNE-GRID.

Finally the local finite elements in DUNE-LOCALFUNCTIONS contain a class for the `LocalCoefficients`

$$\hat{\mathscr{C}}_R = \big((\hat{e}_0, l_0), \ldots, (\hat{e}_{N_R-1}, l_{N_R-1})\big) \subset \bigcup_{c=0}^{d} \hat{E}^c \times \mathbb{N}.$$

For $i = 0, \ldots, N_R - 1$, each nodal variable $\lambda_i \in \boldsymbol{\Lambda}_R$ is assigned a subentity $\hat{e}_i \in \hat{E}_R^{c_i}$ such that the value of $\lambda_i(v)$ only depends on the restriction of v to $\hat{e}_i$. The l_i provide a consecutive numbering of all basis functions which are assigned to the same subentity. In most cases the keys are easily implemented together with the nodal variables.

3 Examples

In the following we will present a few examples showing how the abstract concept described so far can be used to implement standard finite element spaces of arbitrary order and any reference element in the set $\hat{\mathscr{R}}^d$ with $d \geq 1$.

Example 1 (Lagrange shape functions). General Lagrange type elements are defined through point evaluation in some point set $P_L \subset R$. These points are distributed on the subentities of R to allow for the construction of globally continuous discrete function spaces V_h.

Pre basis:	$\hat{\boldsymbol{B}}_R = \mathscr{M}_R^k$
Functionals:	$\lambda_p(u) = u(p) \quad \text{for } p \in P_L$

In the 2.1 release of DUNE-LOCALFUNCTIONS an equidistributed point set for all reference elements and a Lobatto type point set for cubes and simplices is implemented. The latter is based on ideas from [8]. Fig. 2 shows the two different point sets and resulting basis functions for polynomial degree 5.

In release 2.1 of DUNE-LOCALFUNCTIONS arbitrary order Raviart-Thomas elements are available for simplex reference elements. For the construction of this space on general cube type elements see for example [1].

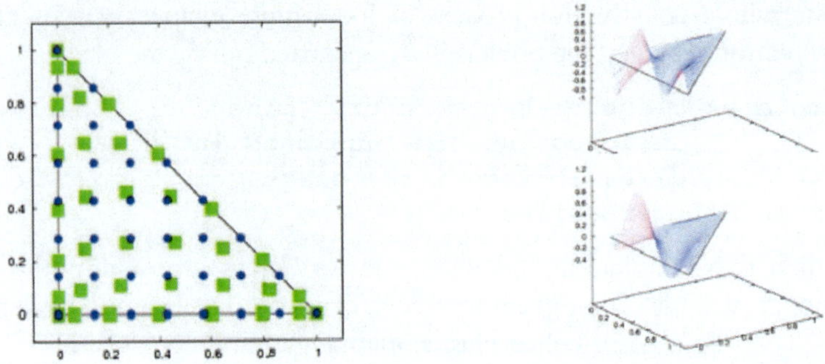

Fig. 2 Points chosen equidistantly (circles) and clustered nearer to the vertices (squares), leading to reduced oscillatory basis functions (bottom right) compared to using equidistant points (top right). The polynomial order in these figures in $k = 5$.

Example 2 (Raviart-Thomas elements using local L^2 projection). Let $B_q(E)$ be a basis of $\mathbb{P}^q(E)$ for any set E and polynomial order q.

Pre basis: $$\hat{\boldsymbol{B}}_R = (\mathscr{M}_R^k)^d + x.\bar{\mathscr{M}}_R^k$$

Functionals:
$$\lambda_{\hat{e},p}(u) := \int_{\hat{e}} u \cdot n_{\hat{e}}\, p \quad \text{for all } \hat{e} \in \hat{E}_R^1 \text{ and } p \in B_k(\hat{e})$$
$$\lambda_{R,p,j}(u) := \int_R u \cdot e_j\, p \quad \text{for all } p \in B_{k-1}(R) \text{ and } j = 1,\ldots,d$$

Note that the pre basis can be written in the form $B.\mathscr{M}_R^{k+1}$ for some matrix B. The implementation of the functionals requires high order quadrature rules on the reference elements, which can be constructed using the recursive definition of the reference elements. For the basis $B_k(\hat{e})$ with $\hat{e} \in \hat{E}_R^1$ we could use the monomial basis, i.e., $B_k(\hat{e}) = \{p \circ i_{\hat{e}}^{-1} : p \in \mathscr{M}_{R_{\hat{e}}}^k\}$. It turns out, though, that using other basis functions to define the nodal variables increases stability (see Section 4).

A second approach for constructing the basis functions is to use functionals based on pointwise evaluation.

Example 3 (Raviart-Thomas elements using point evaluation). Let R be a simplex and let P_L denote a set of points used to construct a Lagrange space of order $k+d$ on R. For $p \in P_L$ let us denote by $\hat{e}(p)$ the subentity of R with maximal codimension containing p and define $P_L^c = \{p \in P_L : \hat{e}(p) \in \hat{E}_R^c\}$.

Pre basis: $$\hat{\boldsymbol{B}}_R = (\mathscr{M}_R^k)^d + x.\bar{\mathscr{M}}_R^k$$

Functionals:
$$\lambda_p(u) := u(p) \cdot n_{\hat{e}(p)} \quad \text{for all } p \in P_L^1$$
$$\lambda_{p,j}(u) := u(p) \cdot e_j \quad \text{for all } p \in P_L^0 \text{ and } j = 1,\ldots,d$$

Any set of functionals used to construct a Lagrange basis space can be used here. In DUNE-LOCALFUNCTIONS the subentity information required can be obtained from the `LocalCoefficients`.

Example 4 (L^2-orthonormal basis functions). Let $S^d = P^{\circ\cdots\circ} \in \hat{\mathscr{R}}^d$ denote the reference element of the d-dimensional simplex. To obtain a set of L^2-orthonormal basis functions on $R \in \hat{\mathscr{R}}^d$, we construct the matrix A_R from a bilinear form a_R as in (2):

Pre basis:	$\hat{\mathbf{B}}_R = \mathscr{M}^k_{S^d}$
Bilinear form:	$a_R(u,v) = \int_R uv$

Based on the recursive definition of the reference elements it is straightforward to define quadrature rules for any order on the reference element R which can be used to evaluate a_R.

4 Numerical Examples

The construction of a global discrete function space V_h from the basis sets $\hat{\mathscr{B}}_R$ is not trivial and is outside of the scope of this paper. For the examples shown here, we only use conforming, *twist-free* grids $\mathscr{T}_h$ for which the construction is straightforward but still rather technical.

Definition 3 (Twist-free grids). For a subentity $e \in E^c_T$ of an element $T \in \mathscr{T}_h$ we define the *twist mapping*

$$t_{T,e} = F_e^{-1} \circ F_T \circ i_{\hat{e}}$$

mapping the reference element $\hat{e} \in \hat{\mathscr{R}}^{d-c}$ of e into itself. We say that a grid $\mathscr{T}_h$ is *twist-free*, if $t_{T,e} = \mathbb{I}$ for all $T \in \mathscr{T}_h$ and $e \in E^c_T$, where $\mathbb{I}$ denotes the identity.

Examples of twist-free grids are the structured grids `SGrid` and `YaspGrid`. But the unstructured, simplicial grid `ParallelSimplexGrid` [9], which we use for the simulations shown here, also has this property due to the special construction of our reference elements.

The local face DoFs of a Raviart-Thomas spaces are associated with a basis function pointing in the direction of the outside normal. To store them, the global DoF has to be negated for one of the two adjacent elements; we arbitrarily choose the element with the larger index to store use the negated global DoF.

To demonstrate the effectiveness of our approach, we compute high order approximations to the gradient of the function

$$u(x,y,z) = \sin\left(32\pi(x-x^2)(y^3-y^2)\right)$$

into different versions of the Raviart-Thomas space. Note that the function u does not depend on z and we use it for both 2D and 3D computations.

Unless otherwise noted, all computations were performed using standard double precision during the evaluation phase. For the construction of the basis functions the high precision floating type `ampf<512>` from [3] was used. Note that in most cases a lower precision would have sufficed.

We have implemented the functionals for both RT spaces described in Examples 2 and 3. Depending on the choice of the polynomial bases used for the local L^2-projection or the point set used for the point evaluation, a wide range of different basis sets can be constructed. We will test three versions:

L^2-ONB: We use the orthonormal basis described in Example 4 to compute the local L^2-projection.

L^2-Lob: We use Lagrange type basis functions constructed with the Lobatto point set as described in Example 1 for the local L^2-projection.

P-Lob: We use the Lobatto point set describe in Example 1 for the pointwise evaluation.

In addition to the Lobatto point set, we have also tested the equidistant point set but found the results to be less satisfactory.

Apart from computing the interpolation based on the functionals (termed *functional-interpolation* in the following), we also assemble the mass matrix and perform an *L^2-projection* of ∇u into the RT space. To solve the resulting system an unpreconditioned CG method is used. Note that for the *functional-interpolation* the two basis sets based on the local L^2-projection produce the same discrete function while the P-Lob method results in a different one. The L^2-projection should lead to the same discrete function, independent of the basis used to describe the RT space and this is what we observed. Therefore we do not distinguish the different spaces when showing L^2-projection errors. Although the discrete functions produced are the same, the condition number of the mass matrix and, therefore, the number of iterations in the linear solver is expected to differ. We demonstrate this by showing results for the condition number based on the L^2 norm or by showing the number of iterations needed to reduce the residual to $1e-14$ in the CG method.

We start off with showing exponential convergence of the interpolation error when increasing the polynomial degree on a fixed grid. The results displayed on the left of Fig. 3 show the error up to $k=15$. For $k>12$, higher floating point precision was required (we used `ampf<256>`, though lower precision would possibly have sufficed). The right plot in Fig. 3 shows exponential convergence in the polynomial degree k in 3D, comparing point evaluation and L^2-interpolation functionals and also L^2-projection error. In this example L^2-projection clearly leads to the best approximation but very closely followed by the functional interpolation based on L^2 projection. Functional interpolation based on the point evaluation in the Lobatto

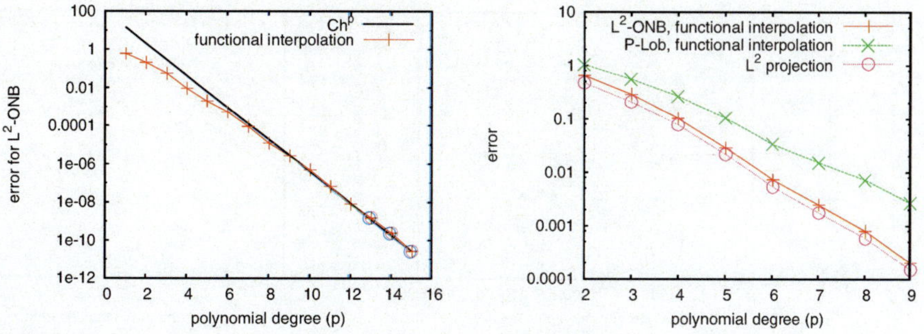

Fig. 3 Exponential convergence in p of interpolation error. (left): 2D Raviart-Thomas spaces using 123 triangles - last three computations with higher order precision. (right): 3D spaces including L^2 (on 233 tetrahedra).

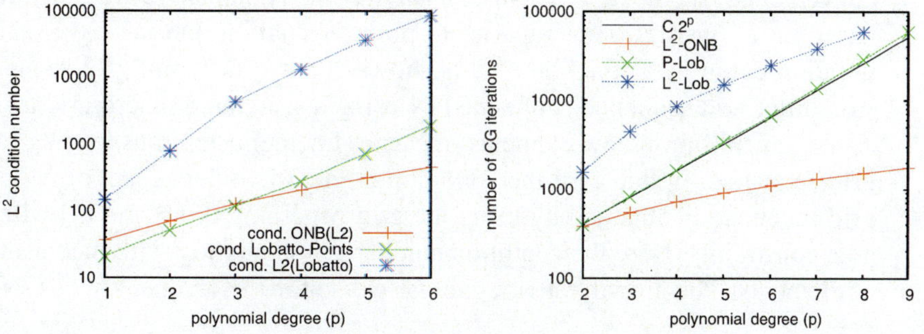

Fig. 4 Study of p refinement: Condition number in 2D on 28 triangles (left) and of number of unpreconditioned CG iterations in 3D on 233 tetrahedra (right).

points leads to an approximation with more than one order of magnitude higher error - an equidistant point set fares even worse.

Next we study the conditioning of the different basis sets in Fig. 4. On the left we present the condition number of the mass matrix based on the L^2-norm as a function of k for our different basis sets in 2D. On the right the number of CG iterations required for inverting the mass matrix in 3D is plotted. It is clear from these results that L^2-ONB is superior to the other approaches but also that functionals based on L^2 projection do not in general lead to better conditioned basis sets than using point evaluation. In fact for small polynomial degrees P-Lob is slightly better conditioned, but the condition number seems to grow exponentially with k while for L^2-ONB the growth rate is much smaller. The ansatz L^2-Lob leads to a significantly higher condition number. Compared to L^2-ONB the number of iterations increases by a factor of 23 when L^2-Lob is used. Using a standard Lagrange basis on an equidistant point set fares even worse, increasing the iteration count by further 50%.

Finally, we show some results obtained under h refinement in 3D for fixed polynomial degree $k = 3$. Fig. 5(left) demonstrates that all spaces show the expected

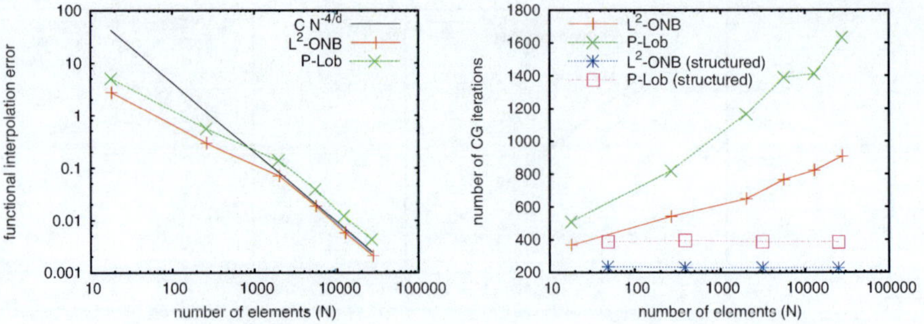

Fig. 5 Study of h convergence for $k = 3$ for dimension $d = 3$. (left): interpolation error. (right): number of CG iterations.

convergence rate of $k+1$ in h since $h = O(N^{-\frac{1}{3}})$. Again the functional interpolation based on L^2 projection is superior to point evaluation. On the right of Fig. 5 the iteration count as a function of h is shown for two different grid sequences. The structured grid sequence is obtained by using a structured macro grid and dividing each element into 8 new elements in each refinement step thus leading to a nested grid sequence. In this case the condition number is independent of h. The second grid sequence is obtained by using the grid generator `Tetgen` [11] with different area constraints. Here the condition number grows slightly with increasing number of elements. The growth rate is again far worse for P-Lob than for L^2-ONB.

5 Conclusion

In this paper we demonstrated that generic construction of very high order basis functions is feasible for a wide range of different finite element spaces. Our construction is based on their mathematical description through functionals. By using high precision floating point classes during the construction phase, high accuracy is achieved and for very high order our implementation allows to use high precision also during the evaluation phase. Since caching of basis function values is an integral component of an efficient high order finite element implementation, both construction and evaluation can be viewed as part of the start up phase of a simulation. Therefore the loss of efficiency due to high precision floating points will not lead to a deterioration of the efficiency of the overall simulation.

It is well known that the correct choice of functionals and even the ways these are realized, e.g., which set of basis functions or which point set is used, strongly influence the stability and accuracy of the approximation. In our numerical tests we demonstrated how our approach allows to easily test many different versions of a given finite element space. In the situation shown here, only the different sets of functionals used to define the spaces had to be implemented. In fact, due to a wide set of utility classes, it is often the case that only the implementation of the functionals is required to obtain the full set of basis functions.

As pointed out above efficiency and stability is in most cases not an issue. Nevertheless, it might be possible to improve the implementation by replacing the monomial basis function set by a set based on Legendre/Jacobi polynomials. The evaluation of these polynomials is in general more stable. Furthermore our implementation allows the output of the basis set in different formats. This is made possible by replacing the field type in the finite-element implementation by a special `MultiIndex` field type. An example of this is shown in Fig. 6. This feature can be used to employ some computer algebra system for automatic optimized code generation. This would also reduce the high compilation time caused by the extreme use of template meta programming.

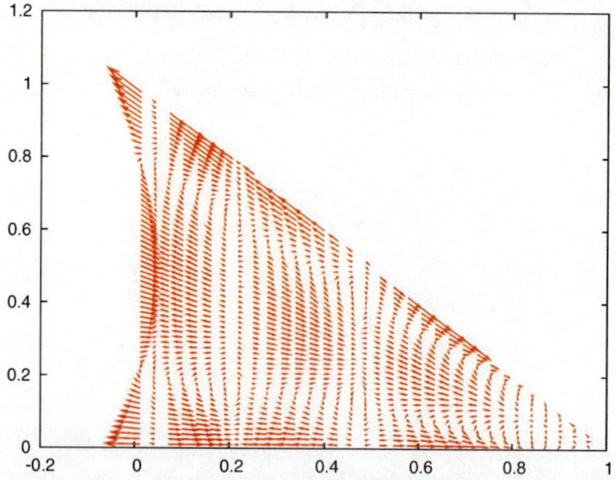

$$\begin{gathered}\varphi_8(x,y) = (-0.2236067977499786E1 + 0.7634574951749300E2x + 0.1341640786499875E2y - \\ 0.6203491446149419E3x^2 - 0.4446580892399579E3xy - 0.1341640786499875E2y^2 + \\ 0.2022044328224809E4x^3 + 0.3277436778449686E4x^2y + 0.4312416813749587E3xy^2 - \\ 0.3109731680137207E4x^4 - 0.8854829190899152E4x^3y - 0.2846195097074728E4x^2y^2 + \\ 0.2266414614337293E4x^5 + 0.9803560889924061E4x^4y + 0.6008634093824425E4x^3y^2 - \\ 0.6324877993499395E3x^6 - 0.3794926796099637E4x^5y - 0.3794926796099637E4x^4y^2, \\ 0.4152697672499613E1y - 0.1277753129999878E3xy - 0.1916629694999824E2y^2 + \\ 0.8567334736649209E3x^2y + 0.5864886866699453E3xy^2 + 0.1724966725499837E2y^3 - \\ 0.2050793773649813E4x^3y - 0.3622430123549653E4x^2y^2 - 0.5174900176499505E3xy^3 + \\ 0.1950170714662316E4x^4y + 0.6957365792849334E4x^3y^2 + 0.2846195097074728E4x^2y^3 - \\ 0.6324877993499395E3x^5y - 0.3794926796099637E4x^4y^2 - 0.3794926796099637E4x^3y^3)\end{gathered}$$

Fig. 6 Graph and latex code generated for a fifth order basis function of a 2D Raviart-Thomas space.

References

1. Arnold, D.N., Boffi, D., Falk, R.S.: Quadrilateral $H(\text{div})$ finite elements. SIAM J. Numer. Anal. 42(6), 2429–2451 (2005)
2. Bastian, P., Blatt, M., Dedner, A., Engwer, C., Klöfkorn, R., Ohlberger, M., Sander, O.: A generic grid interface for parallel and adaptive scientific computing. I: Abstract framework. Computing 82(2-3), 103–119 (2008)

3. Bochkanov, S., Bystritsky, V.: Alglib, http://www.alglib.net
4. Brezzi, F., Fortin, M.: Mixed and Hybrid Finite Elements. Springer (1991)
5. Ciarlet, P.: The Finite Element Method for Elliptic Problems. 2nd Printing. North-Holland (1987)
6. Dedner, A., Klöfkorn, R., Nolte, M., Ohlberger, M.: A generic interface for parallel and adaptive scientific computing: Abstraction principles and the DUNE-FEM module. Computing 89(1) (2010)
7. GMP: The gnu multiple precision arithmetic library, http://www.gmplib.org
8. Luo, H., Pozrikidis, C.: A Lobatto interpolation grid in the tetrahedron. IMA J. Appl. Math. 71(2), 298–313 (2006)
9. Nolte, M.: The dune-psg module, http://dune.mathematik.uni-freiburg.de/grids/dune-psg/
10. DUNE team: The DUNE-LOCALFUNCTIONS manual, http://www.dune-project.org/doc/localfunctions/dune-localfunctions-manual.pdf
11. TetGen: A quality tetrahedral mesh generator, http://tetgen.berlios.de

DUNE-FEM: A General Purpose Discretization Toolbox for Parallel and Adaptive Scientific Computing

Andreas Dedner, Robert Klöfkorn, Martin Nolte, and Mario Ohlberger

Abstract. DUNE-FEM is a free discretization toolbox for parallel and adaptive scientific computing based on DUNE. The implementation of discretization schemes such as finite elements, finite volumes or discontinuous Galerkin schemes is based on abstractions that are very close to the mathematical description of the underlying methods. In this contribution we will give a compact overview on the design and abstraction principles of DUNE-FEM and demonstrate its wide range of applicability in numerical experiments ranging from the solution of flow processes on surfaces to parallel and adaptive fluid flow in three space dimensions. A more detailed presentation of the abstraction principles is given in [Dedner et al. A generic interface for parallel and adaptive discretization scheme: abstraction principles and the Dune-FEM module. Computing 90 (2010), no. 3-4, 165–196]. In the whole design of DUNE-FEM efficiency was a main concern. In this paper we will give some indication to what degree the generic programming principals used in DUNE-FEM can lead to the generation of efficient code.

1 Introduction

The DUNE-FEM module [13, 6] is a free and open source discretization toolbox for the numerical solution of systems of partial differential equations. It is realized in C++ and based on the DUNE core modules [14], in particular on the

Andreas Dedner
Mathematics Institute, University of Warwick, Coventry CV4 7AL, UK
e-mail: A.S.Dedner@warwick.ac.uk

Robert Klöfkorn
Institut für Angewandte Analysis und Numerische Simulation
University of Stuttgart Pfaffenwaldring 57 D-70569 Stuttgart
e-mail: robertk@mathematik.uni-stuttgart.de

Martin Nolte
Abteilung für Angewandte Mathematik, Universität Freiburg,
Hermann-Herder-Str. 10, 79104 Freiburg i. Br., Germany,
e-mail: nolte@mathematik.uni-freiburg.de

Mario Ohlberger
Institute of Computational and Applied Mathematics, University of Muenster, Einsteinstr. 62, 48149 Muenster, Germany
e-mail: ohlberger@uni-muenster.de

DUNE-GRID library [2, 1] that specifies interface classes to handle arbitrary structured, unstructured, adaptive, and parallel computational grids. DUNE-FEM offers interface classes for discrete functions and discrete operators and thus offers a natural framework for the implementation of discretization schemes. In particular, the framework covers finite element, finite volume and discontinuous Galerkin methods for stationary and time dependent partial differential equations. It supports the handling of sequences of varying discrete function spaces that reflects local mesh adaptivity and parallelization with dynamic load balancing of the corresponding discrete function data. General concepts are realized for the discretization of coupled systems, or for the integration in time of space discretized evolution systems. Other attempts have been undertaken to support the implementation of discretization schemes based on the DUNE core modules. In particular, we refer to DUNE-PDELAB [3] where similar design principles are realized.

The design of DUNE-FEM is focused around the efficient implementation especially of matrix free schemes for general non-linear evolution equations on parallel, adaptive grids. To achieve this aim the organization of the storage for the degrees of freedom needs to be implemented carefully because the redistribution of date takes up a significant percentage of the overall runtime. Furthermore, the computational cost of the grid traversal, evaluation of the geometric information, and base function evaluation in higher order scheme, needs to be reduced as much as possible. To this end, special index sets, caching of basis functions at quadrature points and automatic code generation for evaluating local functions at quadrature points is included in DUNE-FEM. To demonstrate the effectiveness of these techniques, we will present some measurements of the parallel efficiency and the of the FLOPS. A further study of these aspects is presented in [4] where we study the efficiency of DUNE-FEM for solving meteorological problems, comparing it with the finite-difference code COSMO used for the operational weather forecast at the Deutscher Wetterdienst.

The rest of the paper is organized as follows. In Section 2 we shortly review the principal design guidelines of DUNE-FEM and comment on the basic implemented interface classes. In the following Section 3, practical aspects of the implementation of discretization schemes in DUNE-FEM are discussed. Finally, in Sections 4 and 5, several numerical experiments are discussed ranging from the solution of flow processes on surfaces to parallel and adaptive fluid flow in three space dimensions.

2 Abstraction Principles and Interface Classes

A principal guideline for the design of DUNE-FEM is to follow as closely as possible the abstract mathematical notion of discretization schemes for systems of partial differential equations, while maintaining at the same time efficiency when it comes to the solution of the underlying systems of algebraic equations. This aim is achieved through the extensive use of template based programming techniques in C++ that allow to separate functionality from data structures without loss of efficiency. In the following, we first describe the mathematical framework used as a guideline for the definition of interface classes forming the backbone for the implementation of a large class of discretization schemes.

2.1 Mathematical Framework

Let U, V denote some function spaces. We start our considerations for stationary problems that we consider in the following abstract form. Find a solution $u \in U$ of

$$\mathscr{L}(u) = F \quad \text{in } V', \tag{1}$$

where $\mathscr{L} : \mathscr{D}(\mathscr{L}) \subset U \to V'$ denotes a differential operator, mapping functions from the domain $\mathscr{D}(\mathscr{L})$ of the the operator $\mathscr{L}$ into the dual space V' of V and $F \in V'$ is a suitable right hand side.

The most famous example is Poisson's equation, $-\Delta u = f$, in some domain Ω with a Dirichlet boundary condition $v = 0$ on $\partial\Omega$. In this case one might consider $\mathscr{D}(\mathscr{L}) = U = V = H^1_0(\Omega)$ and the weak solution is defined by (1) with

$$\mathscr{L}(u)[v] := \int_\Omega \nabla u \cdot \nabla v, \qquad F[v] := \int_\Omega f v.$$

Starting from this abstract definition of stationary problems, a large class of discretization schemes can be written in the abstract form

$$\mathscr{L}_h(u_h) = F_h \quad \text{in } V'_h, \tag{2}$$

where now $\mathscr{L}_h : U_h \to V'_h$ is a discrete operator and $F_h \in V'_h$ is a discrete right hand side. Of course, the discrete function space U_h and V_h should be chosen such that $U_h \to \mathscr{D}(\mathscr{L}) \subset U$ and $V_h \to V$ as $h \to 0$.

The abstract form (2) can reflect standard finite element discretizations, if U_h, V_h are globally continuous, piecewise polynomial subspaces of U, V. But also Petrov Galerkin discretizations, discontinuous Galerkin approximations, or finite volume schemes, and even schemes based on Fourier series can be represented in this abstract form for suitable choices of discrete operators and function spaces. For suitable bases $\Phi_h := \{\phi_i | i = 1, \ldots N := \dim(U_h)\}$ of U_h and $\Psi_h := \{\psi_i | i = 1, \ldots M := \dim(V_h)\}$ of V_h, the discretization scheme (2) is equivalent to the algebraic system

$$\mathbf{L}(\mathbf{u}) = \mathbf{F} \tag{3}$$

where $\mathbf{u} \in \mathbb{R}^\mathbf{N}$ denotes the coefficient vector of u_h with respect to the basis Φ_h, i.e., $u_h = \sum_{i=1}^N \mathbf{u}_i \Phi_i$, and $\mathbf{L}, \mathbf{F}$ are defined through

$$\mathbf{L}(\mathbf{u})_i := \mathscr{L}_h(u_h)[\psi_i], \qquad \mathbf{F}_i := F_h[\psi_i] \qquad \text{for } i = 1, \ldots M.$$

Evolution equations of the general form

$$\partial_t u(\cdot, t) = \mathscr{L}(u(\cdot, t))$$

with a differential space operator $\mathscr{L} : \mathscr{D}(\mathscr{L}) \subset U \to V'$ can be treated by combining the framework for the stationary case with solvers for ordinary differential equations using, for example, the method of lines approach.

From the above observations we conclude that an abstract definition of discretization schemes can be obtained through a proper definition of discrete function spaces and discrete operators.

2.2 Abstraction and Interface Classes

In [6] we deduced from the general mathematical framework given in Subsection 2.1 an abstraction for (localized) discrete functions, discrete spatial operators, related inverse operators and corresponding generalizations for sequences of function spaces associated with adaptive mesh refinement. In addition, abstractions for parallelization, data exchange and dynamic load balancing were introduced (see [6, Sections 2.4, 2.5]). In the following we restrict ourselves to a more practical introduction of the major C++ interface classes in DUNE-FEM and refer the interested reader to [6] for a more detailed discussion of the abstraction.

We start with a description of interface classes representing (discrete) functions, where a function space V is understood as a set of mappings $V := \{u : \mathbb{K}_D^d \to \mathbb{K}_R^n\}$. A localized discrete function space V_h with finite dimension m is a subset of a function space with the property that the functions are defined locally on elements e of the underlying computational grid $\mathscr{T}$. If $\hat{e}$ denotes the reference element of e and F_e the mapping $F_e : \hat{e} \to e$, the local basis function set $V_{\hat{e}}$ is given through $V_{\hat{e}} := \{\varphi_1, \ldots, \varphi_{\dim(V_{\hat{e}})}\}$. Hence, a localized discrete function space V_h can be expressed as

$$V_h := \Big\{ u_h \in V : u_e := u_h|_e := \sum_{\varphi \in V_{\hat{e}}} g(u_{e,\varphi})\ \varphi \circ F_e^{-1}, \text{ for all } e \in \mathscr{T} \Big\}$$

where $g(u_{e,\varphi})$ denotes the global degree of freedom (DoF) corresponding to the local DOF $u_{e,\varphi}$. The major interface classes in DUNE-FEM that represent these concepts are

1. `FunctionSpace<DomainField,RangeField,dimD,dimR>`: represents a function space $V \subset \{v : \mathbb{K}_D^d \to \mathbb{K}_R^n\}$ with $\mathbb{K}_D =$ DomainField, $\mathbb{K}_R =$ RangeField, and $d =$ dimD, $n =$ dimR.
2. `Function<FunctionSpace>`: represents a function $u \in V$. The most important method is `evaluate(x, ret)`: $ret = u(x)$, for $u \in V$.
3. `DiscreteFunctionSpace<FunctionSpaceTraits>`: represents a discrete function space V_h. It is parametrized by the type of the function space V such that $V_h \subset V$, the type of the computational grid $\mathscr{T}$ and the type of the basis function set $V_{\hat{e}}$. The class provides an iterator to access the entities e of $\mathscr{T}$. The method `mapToGlobal(e, nLocal)` returns the global number of the degree of freedom with local number `nLocal` on entity `e`, and `getBaseFunctionSet(e)` returns the local base function set $V_{\hat{e}}$ of `e`.
4. `DiscreteFunction<Traits>`: a discrete function is parametrized with the type of the discrete function space V_h it belongs to and with the type of its local

functions u_e on the entities e. To access local functions, it provides the method `lf = localFunction(e)` that returns the local function `lf` of entity `e`.

5. `LocalFunction<Traits>`: represents a local function u_e. Providing evaluation methods like `evaluate(x,ret)` and `jacobian(x,ret)`, evaluating $ret = u_e(F_e(x))$ and $ret = Du_e(F_e(x))$, respectively.

As depicted in the mathematical description in Sec. 2.1, the second major ingredient for the implementation of discretization schemes are (discrete) operators, and corresponding inverse operators that represent the linear or non-linear solvers for the resulting algebraic systems. The basic interface classes in DUNE-FEM are

1. `Operator<DomainFunction, RangeFunction>`: represents an operator $\mathscr{L}$, that maps from a domain function space to a range function space.
2. `DiscreteOperator<LocalOperator,DFDomain,DFRange>` represents a discrete operator $\mathscr{L}_h$ that maps from a discrete function space (`DFDomain`) into another discrete function space (`DFRange`). In addition the type of the local Operator L_e, acting on an entity e is given. To apply the discrete operator, the C++ class method `operator()(arg,dest)` has to be called.
3. `InverseOperator<DiscreteFunction,DiscreteOperator>`: represents a numerical solver to invert a given `DiscreteOperator`. The inverse operator is applied to a `DiscreteFunction`. Examples for implemented inverse operators are `CGInverseOperator`, `OEMCGOp`, `OEMBICGSTABOp`, `OEMGMRESOp` or solvers from DUNE-ISTL. As a direct solver UMFPACK [6] can be used via `UMFPACKOp`. Non-linear solvers are also available based on various forms of the Newton method, including matrix free methods.

To solve time dependent problems using the method of lines, efficient solvers for large systems of ODEs are required. The basic interface class in DUNE-FEM is `OdeSolverInterface<DiscreteFunctionSpace>`. Implementations include different types of implicit, semi-implicit, and explicit Runge-Kutta methods.

In addition to the interface classes required for solving partial differential equations, DUNE-FEM includes a number of classes for input/output (including parameter parsing, check-pointing, and visualization) and for facilitation of the use of grid adaptivity and parallelization.

3 Practical Aspects and Code Snippets

In this Section we demonstrate the use of the classes defined in the DUNE-FEM module. As an example an L^2-projection of a given analytical function f and the projection error are computed. The discrete function u in some discrete function space V_h is defined by the equation

$$\int_\Omega u\varphi = \int_\Omega f\varphi \quad \text{for all } \varphi \in V_h .$$

The first code snippet (see Listing 1) shows how discrete function spaces, functions and a linear operator are constructed. The construction process is based on a

`GridPart` which is used in DUNE-FEM instead of the concept of `GridView` used in DUNE-GRID. The main difference is the construction process of these views: as can be seen in Listing 1, `GridParts` are not obtained from the grid but are initialized by passing a grid object in the constructor. After choosing a `GridPart`, an instance of a Lagrange function u on the leaf grid is created. The function maps to $\mathbb{R}^r$, i.e., is vector valued if the integer constant $r > 1$. The order of the Lagrange space is p given in the template argument list. Finally a sparse row matrix is constructed. The sparsity pattern for the matrix can be initialized through a template argument depending on the discrete function space used. Different implementations are available, e.g., with bindings for DUNE-ISTL.

The required type for the local function u_e and the local operator $M_{e,e}$ can be easily accessed through the types of the discrete function space and the linear operator. Other types, e.g., the iterator over the grid or the type of the local base function set $V_{\hat{e}}$ are also provided by the discrete function space. The type of the function f, which is to be projected, should be derived from the `Fem::Function` interface class and has to provide a method `evaluate(x,y)` which implements $y = f(x)$.

Listing 1 Constructing discrete spaces and functions

```
// type of the grid, where HGridType is the DUNE-GRID type of ℋ.
typedef AdaptiveLeafGridPart<HGridType, InteriorBorder_Partition>
        GridPartType;
// type of the function space
typedef FunctionSpace<double, double, HGridType::dimensionworld, r>
        FunctionSpaceType;
// type of the discrete function space
typedef LagrangeDiscreteFunctionSpace
        <FunctionSpaceType, GridPartType, p>
        DiscreteSpaceType;
// type of the discrete function
typedef AdaptiveDiscreteFunction<DiscreteSpaceType>
        DiscreteFunctionType;

// type of the linear operator for the mass matrix
typedef SparseRowMatrixOperator
        <DiscreteFunctionType, DiscreteFunctionType,
         SparsityPatternTraits<DiscreteSpaceType> >
        MatrixType;

// extract required types
typedef DiscreteSpaceType::RangeType RangeType;
typedef DiscreteSpaceType::IteratorType IteratorType;
typedef DiscreteSpaceType::BaseFunctionSetType BaseFunctionSetType;

typedef DiscreteFunctionType::LocalFunctionType LocalFunctionType;
typedef MatrixType::LocalMatrixType LocalMatrixType;

// construct leaf grid 𝒯_h from a hierarchical grid ℋ
GridPartType grid( hgrid );
// construct lagrange discrete function space V_h
DiscreteSpaceType space( grid );

// create the analytical function f to project
RightHandSideFunction<FunctionSpaceType> f;
// create the solution u
DiscreteFunctionType u( "solution", space );
// create the functional b and the linear operator M
DiscreteFunctionType b( "functional", space );
MatrixType M( "mass_matrix", space, space );
```

In Listing 2, we perform a grid walk-through to assemble the mass matrix M with entries $M_{ij} = \int_\Omega \varphi_i \varphi_j$, and the functional b on the right hand side, given by $b_j = \int_\Omega f \varphi_j$. As usual in the finite-element framework, the assembly is performed element wise. To facility this step, special *axpy like* methods are provided on the local functions and matrices:

- **void** axpy (**const** PointType &x, **const** RangeType &s);
  $b_{e,\varphi} \mathrel{+}= s \cdot \varphi(x)$ for all $\varphi \in V_{\hat{e}}$.
- **void** axpy (**const** PointType &x, **const** JacobianRangeType &ds);
  $b_{e,\varphi} \mathrel{+}= ds : \mathrm{D}\varphi(x)$ for all $\varphi \in V_{\hat{e}}$.
- **void** axpy (**const** PointType &x, **const** RangeType &s, **const** JacobianRangeType &ds);
  Equivalent to `axpy( x, s ); axpy( x, ds );`

Listing 2 Assembling mass matrix and right hand side

```
// initialization of M and b
M.reserve();
M.clear();
b.clear();

// create some temporary storage for the values of the local base functions V_ê_e
std::vector<RangeType> values;

// walk over the grid T_h
const IteratorType end = space.end();
for( IteratorType it = space.begin(); it != end; ++it )
{
  const EntityType &entity = *it;
  const GeometryType &geometry = entity.geometry();

  // obtain local views to the functional
  LocalFunctionType bLocal = b.localFunction( entity );
  // obtain local operator M_e,e
  LocalMatrixType MLocal = M.localMatrix( entity, entity );

  // obtain the local base function set V_ê_e
  const BaseFunctionSetType &bFuncSet = space.baseFunctionSet(entity);
  const unsigned int numBaseFunctions = bFuncSet.numBaseFunctions();
  values.resize( numBaseFunctions );

  // compute the integrals ∫_e φ_i φ_j and ∫_e f φ_i using a quadrature with base function caching
  typedef CachingQuadrature<GridPartType, 0> QuadratureType;
  QuadratureType quadrature( entity, 2*space.order()+1 );
  const unsigned int nop = quadrature.nop();
  for( unsigned int qp = 0; qp < nop; ++qp )
  {
    const QuadratureType::CoordinateType &x = quadrature.point( qp );
    const double weight = quadrature.weight( qp ) *
                          geometry.integrationElement( x );

    // compute the values of all base functions at quadrature point.
    // These values are precomputed when using a CachingQuadrature
    bFuncSet.evaluateAll( quadrature[ qp ], values );

    // right hand side assembly
    RangeType fValue;
    f.evaluate( geometry.global( x ), fValue );
    rhsLocal.axpy( quadrature[ pt ], f );

    // mass matrix assembly
    for( unsigned int localCol = 0; localCol < numBaseFunctions; ++localCol )
    {
      // get column object and call axpy method
      MLocal.column( localCol ).axpy( values, values[localCol], weight );
    }
  }
}
```

In the following Listing the mass matrix is inverted to produce $u = M^{-1}b$ using a CG based inverse linear operator.

Listing 3 L^2 projection (inversion of mass matrix)

```
// construct the inverse operator M^-1
CGInverseOp<DiscreteFunctionType , MatrixType>
            inverseOperator( M, 1e-10, 1e-10 );
// compute solution
u.clear();
inverseOperator( b, u );
```

Remark 1. Note the use of the class `CachingQuadrature` which provides a caching mechanism for the basis evaluation. This approach provides a significant speedup. There are different implementations of discrete functions available. The `AdaptiveDiscreteFunction` implementation used in this example is optimized for computations where local grid adaptivity is an important factor in the overall run time of the simulation. A further example is the `BlockVector-DiscreteFunction` which is optimized for using the solvers in DUNE-ISTL.

Listing 4 demonstrates how to handle the discretization of evolution equations in DUNE-FEM. Assuming that a discrete spatial operator `DiscreteOperatorType` is given, the construction of an ODE solver and the simple implementation of a time stepping scheme is shown. The time loop is taken care of by a `GridTimeProvider`, which uses the grid communicator to synchronize time steps in a parallel computation. An upper bound for the next time step Δt is set by using the `ProvideTimeStepEstimate` method. Furthermore the spatial operator `spaceOperator` can also provides a method `timeStepEstimate` to reduce the Δt further to maintain stability.

Listing 4 Using a time stepping scheme to solve an evolution equation

```
// initialize the time provider
GridTimeProvider<HGridType> timeProvider( 0, hgrid );
// create space discretization operator
DiscreteOperatorType spaceOperator( order );
// construct ODE solver
typedef ExplicitRungeKuttaSolver<DiscreteFunctionType> ODESolverType;
ODESolverType odeSolver( spaceOperator, timeProvider, order+1 );

// set the initial time step estimate
odeSolver.initialize( u );
// time loop
for( timeProvider.init(); timeProvider.time() < T;
     timeProvider.next() )
{
  timeProvider.provideTimeStepEstimate( maxTimeStep );
  odeSolver.solve( u );
}
```

We conclude our short survey of the module DUNE-FEM by demonstrating the concepts for including adaptivity and parallelization. Using the adaptation manager from DUNE-FEM and the default restriction/prolongation operators on the discrete function space, the few lines of code for adapting a grid and keeping the discrete functions consistent, are given in Listing 5. The simplest approach for data communication is also shown.

Listing 5 Adapting and communicating a discrete function

```
// type of the default restriction and prolongation operator
typedef RestrictProlongDefault<DiscreteFunctionType>
        RestrictProlongType;
// type of the adaptation manager
typedef AdaptationManager<HGridType, RestrictProlongType> i
        AdaptationManagerType;

// create restriction and prolongation operator for u and the adaptation manager
RestrictProlongType uRestrictProlong( u );
AdaptationManagerType
      adaptationManager( hgrid, uRestrictProlong );

// mark grid for refinement and coarsening using some external method mark
mark( hgrid, u );

// adapt the grid with automatic restriction and prolongation of the discrete function u
adaptManager.adapt();

// communicate u using the space's default communication
space.communicate( u );
```

4 PDEs on Surfaces

In this section, we consider the approximation of partial differential equations on surfaces using DUNE-FEM.

The DUNE-GRID interface natively supports the discretization of d-dimensional manifolds embedded into some Euclidian space $\mathbb{R}^w$ with $d \le w$. For notational purposes, let g denote the Riemannian metric on the manifold induced by the inner product on $\mathbb{R}^w$.

As examples, we will consider Poisson's equation and Burger's equation on the two-dimensional sphere $\mathbb{S}^2 \subset \mathbb{R}^3$.

4.1 Poisson's Equation

As a first example, we will show how to solve the Poisson equation

$$-\Delta_g u = f \quad \text{in } \mathbb{S}^2 \tag{4}$$

using DUNE-FEM, where Δ_g denotes the Laplace-Beltrami operator. This problem is well-posed if and only if f has a zero mean value.

We use linear finite elements to approximate the solution of (4) on a grid consisting of affine triangles. Notice that, while higher order ansatz functions are supported in DUNE-FEM, the geometric approximation error due to the piecewise affine element geometry is of order h^2, so that no improvement in the order of convergence can be expected.

Implementations of the DUNE grid interface supporting surface grids include `AlbertaGrid`, `ALUCubeGrid`, and `ALUSimplexGrid` (for ALUGRID version 2.0). In this example we use `AlbertaGrid` with a macro triangulation being the surface of an octahedron; on grid refinement, new vertices are projected to the sphere $\mathbb{S}^2$. The corresponding code can be found in the DUNE-FEM how-to.

For the numerical validation of the scheme, we choose f such that the exact solution is given by $u(x) = \prod_{i=1}^{3} \sin(2\pi x_i)$. Note that the solution is only unique up to a constant. We use an iterative linear solver with $u_h \equiv 0$ as initial guess, so that the approximated solution will retain zero mean.

Convergence results in $L^2(\mathbb{S}^2)$ and $H^1(\mathbb{S}^2)$ are listed in Table 1. The scheme achieves the optimal order of convergence, i.e., 2 with respect to the L^2-norm and 1 with respect to the H^1-norm. The solution on 524 288 triangles is displayed in Figure 1.

Table 1 Results for the approximation of the solution to Poisson's equation on the sphere $\mathbb{S}^2$ by linear finite elements.

triangles	h	CPU-time	L^2-error	L^2-EOC	H^1-error	H^1-EOC
32	$1.89 \cdot 10^{-1}$	$0.00\,s$	$9.10 \cdot 10^{-1}$	—	$8.45 \cdot 10^{0}$	—
128	$9.75 \cdot 10^{-2}$	$0.00\,s$	$1.56 \cdot 10^{0}$	−0.81	$1.11 \cdot 10^{1}$	−0.41
512	$4.90 \cdot 10^{-2}$	$0.00\,s$	$5.96 \cdot 10^{-1}$	1.40	$7.30 \cdot 10^{0}$	0.61
2048	$2.42 \cdot 10^{-2}$	$0.01\,s$	$2.02 \cdot 10^{-1}$	1.54	$4.08 \cdot 10^{0}$	0.83
8192	$1.20 \cdot 10^{-2}$	$0.03\,s$	$5.27 \cdot 10^{-2}$	1.92	$2.07 \cdot 10^{0}$	0.97
32768	$6.01 \cdot 10^{-3}$	$0.26\,s$	$1.32 \cdot 10^{-2}$	1.99	$1.04 \cdot 10^{0}$	1.00
131072	$3.00 \cdot 10^{-3}$	$1.73\,s$	$3.30 \cdot 10^{-3}$	2.00	$5.19 \cdot 10^{-1}$	1.00
524288	$1.50 \cdot 10^{-3}$	$12.90\,s$	$8.25 \cdot 10^{-4}$	2.00	$2.59 \cdot 10^{-1}$	1.00

Fig. 1 Numerical solution to (4) with linear finite elements on 524 288 triangles.

4.2 Burger's Equation

In this example we study a finite-volume scheme for a scalar conservation law on the unit sphere . The code is based on the finite-volume scheme found in the how-to of DUNE-FEM. This problem was studied in detail in [9].

Consider the initial value problem

$$\begin{aligned} \partial_t u + \nabla_g \cdot \Big(\frac{u^2}{2}\mathbf{v}(x)\Big) &= 0 \quad \text{in } \mathbb{S}^2\times]0,T[, \\ u(\cdot,0) &= u_0 \quad \text{in } \mathbb{S}^2 \end{aligned} \tag{5}$$

for a given, divergence-free velocity field $\mathbf{v} : \mathbb{S}^2 \to T\mathbb{S}^2$ and given initial data $u_0 \in L^\infty(\mathbb{S}^2)$. Here, $\nabla_g \cdot$ denotes the divergence operator induced by the Riemannian metric g.

For the numerical experiments, we use a shear flow

$$\mathbf{v}(\varphi,\theta) = a(\theta)\big(\sin\varphi, \cos\varphi, 0\big),$$

where $\varphi \in [0,2\pi[$ and $\theta \in [-\pi,\pi]$ denote the spherical coordinates of a point.

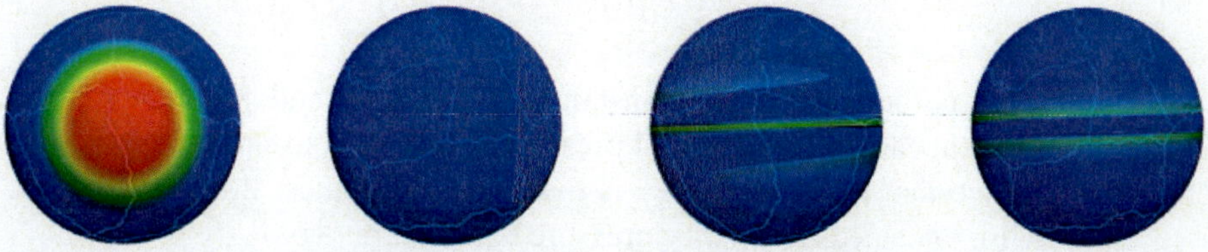

Fig. 2 Numerical solution to (5) with a standard finite volume method on 3 028 288 triangles distributed onto 16 processes. From left to right, the figures show a front view of the initial data, a rear view of the initial data, a front view at $t \approx 0.42$, and a rear view at $t \approx 0.42$. The white lines indicate the process borders.

Fig. 2 shows the initial data and the numerical solution at a later time on a parallel grid with approx. 3 million triangles.

5 The Compressible Navier-Stokes Equations

In this section we give a short description of the Navier-Stokes equation that will be the main point of interest for comparison of the newly proposed method from [5] and other known methods. The nonlinear compressible Navier-Stokes equations can be written in the form

$$\partial_t U + \nabla \cdot \big(F(U) - G(U,\nabla U)\big) = \boldsymbol{S}(U) \qquad \text{in } \boldsymbol{\Omega} \times [0,t_{\text{end}}] \tag{6}$$

where $\Omega \in \mathbb{R}^2$ is a bounded domain (see e.g. [8]). $U = (\rho, \rho\mathbf{v}, \rho e)$ is the unknown vector with the *conservative* variables. ρ is density, $\rho\mathbf{v}$ momentum vector and ρe total energy. In this section we will restrict ourselves to two space dimensions for the sake of simpler notation. Additional variables are $\mathbf{v} = (u, w)$ velocity vector, P pressure and T temperature. The convective and diffusive fluxes are given as

$$F(U) = \begin{bmatrix} \rho u & \rho w \\ \rho u^2 + P & \rho uw \\ \rho uw & \rho w^2 + P \\ u(\rho e + P) & w(\rho e + P) \end{bmatrix}, \; G(U, \nabla U) = \begin{bmatrix} 0 & 0 \\ \tau_{11} & \tau_{12} \\ \tau_{21} & \tau_{22} \\ e_{d,1} + \kappa \partial_x T & e_{d,2} + \kappa \partial_z T \end{bmatrix}$$

with $e_{d,1} = u\tau_{11} + w\tau_{21}$ and $e_{d,2} = u\tau_{12} + w\tau_{22}$. The viscous stress tensor τ for Newtonian fluids is defined as

$$\tau = \begin{bmatrix} (2\mu + \lambda)\partial_x u + \lambda \partial_z w & \mu(\partial_x w + \partial_z u) \\ \mu(\partial_x w + \partial_z u) & \lambda \partial_x u + (2\mu + \lambda)\partial_z w \end{bmatrix}. \tag{7}$$

κ is thermal conductivity coefficient and the term $\kappa \nabla T$ represents the heat flux according to the *Fourier's law*. In the convective and diffusive fluxes we have additional unknowns P and T which are related to U by

$$P = (\gamma - 1)\rho(e - \frac{1}{2}\mathbf{v}^2), \qquad T = P/(\rho R_d) \tag{8}$$

where R_d is the specific gas constant of the dry air and $\gamma = c_P/c_v$ is the ratio of specific heat capacity at constant pressure c_P and at constant volume c_v. Notice that we assume that the total energy is the sum of the internal and the kinetic energy, otherwise one needs to modify pressure equation in (8).

For monoatomic gases we can set $\lambda = -\frac{2}{3}\mu$. Nevertheless this relation is used even for more complicated gases. Finally, we calculate the thermal conductivity κ by using the definition of the Prandtl number which is $\mathrm{Pr} = c_P \mu \kappa^{-1}$. For the calculations we used $\mathrm{Pr} = 1$.

The spatial discretization is carried out by Discontinuous Galerkin (DG) methods. A detailed description of the methods we apply is given in [5]. The discretization includes Interior-Penalty (IP), Local Discontinuous Galerkin (LDG), Bassi-Rebay 2 (BR2), Compact Discontinuous Galerkin (CDG), and Compact Discontinuous Galerkin 2 (CDG2). As time integrator we use implicit-explicit Runge-Kutta (IMEX-RK) methods (e.g. [10]) of different orders, to have better resolution of the advection (treated explicitly) while avoiding severe time step restriction of the diffusion (treated implicitly), which one would have if a fully explicit RK method was used. In [5] a convergence study for the different schemes can be found. In Figure 3 we present a comparison of the implemented DG methods for a test case with analytical exact solution. Figure 3a shows the number of iterations needed by the linear solver called by the Newton solver used to solve each time step. We discover that the CDG2 method needs least iterations in comparison to the other methods. As can be seen in the Figure 3b this leads to the best performance of the CDG2 method when comparing CPU time and L^2-error for basis functions of polynomial order $p = 3$.

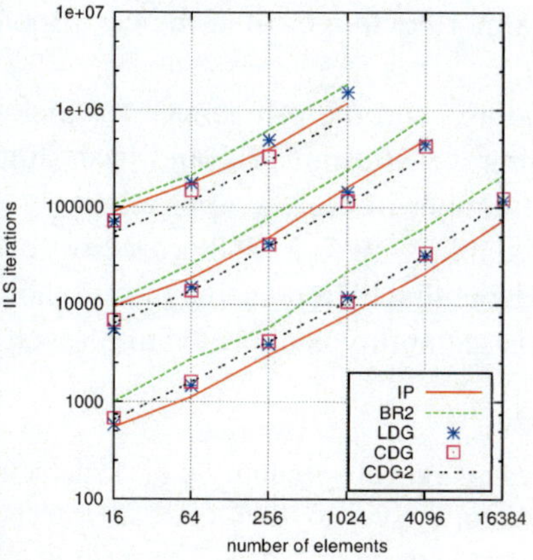

(a) Iterations of the linear solver for $p = 1, 2, 3$.

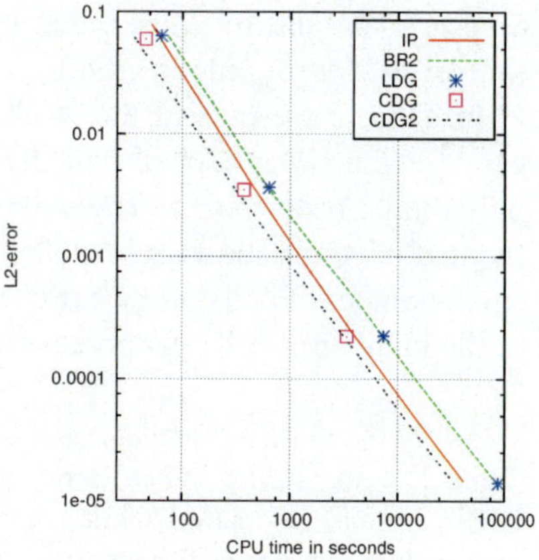

(b) Efficiency for $p = 3$.

Fig. 3 Number of iterations of the linear solver for $p = 1, 2, 3$, respectively, and efficiency for the different DG methods for solution of the Navier-Stokes equations.

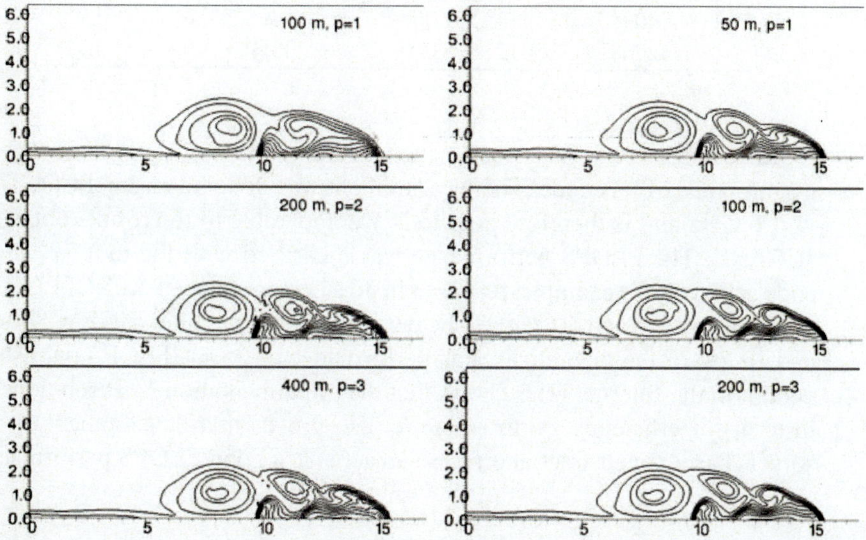

Fig. 4 Potential temperature at different resolutions and orders of the method. We see that the higher order methods can resolve important structures of the solution on coarser grids.

As an application we choose a standard atmospheric test case in [11]. In comparison to the reference solution we observe that the flow is properly resolved, most evident on picture "100m, $p = 2$" where all three vortices are fully developed. Nevertheless, one can spot all vortices at 200m with the fourth order method as well as for the third order at 100m, while for the second order method the flow is not yet resolved at 50m. However, the computational time for the second order method at 50m is 57%, and for third order at 100m 19% longer than the corresponding run time of the fourth order method.

In Table 2 we present the weak scaling[1] and the efficiency[2] for the above test case for the CDG2 scheme in 3D using a Cartesian grid and basis functions of polynomial order $p = 1$. The average number of cells is approximately 3800. We see that the code shows an excellent scalability up to 32768 processors even with a small number of cells on each process. Note that by increasing the polynomial order or thc number of cells per processor this scalability is expected to be even better.

Table 2 Weak scaling and efficiency (speedup/optimal speedup) on the supercomputer JUGENE (Jülich, Germany). While the runs from 8 cores to 4096 cores are executed on one rack of the machine this is not the case for the run on 32768 cores. This leads to a slight drop on speedup and efficiency due to a more expensive communication.

#CPUs	#cells	speedup	efficiency
8	30 375	1.0	1.00
64	243 000	7.9	0.99
512	1 944 000	62.5	0.98
4096	15 552 000	498.2	0.97
32768	125 978 000	3559.3	0.87

Since a slow code scales better than a faster code we briefly discuss the FLOPS performance of this code. The measurement was done on an Intel Core i7 QM 720 @ 1.6 GHz and is therefore not directly comparable to the results obtained on the JUGENE. The FLOPS performance was measured with the tool *likwid* [12]. The code achieves, for example, for $p = 3$ in 3d a performance of 5.05 GFLOPS which is between 33.9%[3] and 40.7%[4] of the performance of the Intel LINPACK Benchmark and 19.7% of the theoretical peak performance of 25.6 GFLOPS published by the vendor in the Internet. This shows that the implementation has been done carefully in terms of efficiency. Furthermore, it also proves that development of numerical solvers based on abstract interfaces can achieve a good FLOPS performance.

References

1. Bastian, P., Blatt, M., Dedner, A., Engwer, C., Klöfkorn, R., Kornhuber, R., Ohlberger, M., Sander, O.: A generic grid interface for parallel and adaptive scientific computing. II: Implementation and tests in dune. Computing 82(2-3), 121–138 (2008)

[1] Change of solution time for a varying number of processors and a fixed problem size per processor.
[2] Efficiency is the measured speedup divided by the optimal speedup.
[3] This result is given by the Intel LINPACK Benchmark based on internal FLOP counting.
[4] LINPACK FLOPS measured with likwid.

2. Bastian, P., Blatt, M., Dedner, A., Engwer, C., Klöfkorn, R., Ohlberger, M., Sander, O.: A generic grid interface for parallel and adaptive scientific computing. I: Abstract framework. Computing 82(2-3), 103–119 (2008)
3. Bastian, P., Heimann, F., Marnach, S.: Generic implementation of finite element methods in the distributed and unified numerics environment (DUNE). Kybernetika (Prague) 46(2), 294–315 (2010)
4. Brdar, S., Baldauf, M., Dedner, A., Klöfkorn, R.: Comparison of dynamical cores for NWP models. Tech. rep(2011); Submitted to Journal of Theoretical and Computational Fluid Dynamic
5. Brdar, S., Dedner, A., Klöfkorn, R.: Compact and stable Discontinuous Galerkin methods for convection-diffusion problems. J. Sci. Comp. 47, 365–389 (2010)
6. Davis, T.A.: Algorithm 832: UMFPACK v4.3—an unsymmetric-pattern multifrontal method. ACM Trans. Math. Softw. 30(2), 196–199 (2004), doi: http://doi.acm.org/10.1145/992200.992206
7. Dedner, A., Klöfkorn, R., Nolte, M., Ohlberger, M.: A generic interface for parallel and adaptive scientific computing: Abstraction principles and the DUNE-FEM module. Computing 89(1) (2010)
8. Feistauer, M., Felcman, J., Straškraba, I.: Mathematical and computational methods for compressible flow. In: Numerical Mathematics and Scientific Computation. Oxford University Press, Oxford (2003)
9. Müller, T.: Erhaltungsgleichungen auf Mannigfaltigkeiten. Wohlgestelltheit, Totalvariationsabschätzungen und Numerik. Diploma thesis, Universität Freiburg (2009)
10. Pareschi, L., Russo, G.: Implicit-explicit Runge-Kutta schemes and applications to hyperbolic systems with relaxation. J. Sci. Comput. 25(1), 129–155 (2005)
11. Straka, J., Wilhelmson, R., Wicker, L., Anderson, J., Droegemeier, K.: Numerical solutions of a non-linear density current: A benckmark solution and comparisons. Int. J. 17, 1–22 (1993)
12. Treibig, J., Hager, G., Wellein, G.: Likwid: A lightweight performance-oriented tool suite for x86 multicore environments. In: Proceedings of PSTI 2010, the First International Workshop on Parallel Software Tools and Tool Infrastructures, San Diego, CA (2010)
13. Website: DUNE-FEM – The FEM Module, http://dune.mathematik.uni-freiburg.de/
14. Website: DUNE – Distributed and Unified Numerics Environment, http://www.dune-project.org

The DUNE-PRISMGRID Module

Christoph Gersbacher

Abstract. In this paper, we describe the design and implementation of the DUNE meta grid module DUNE-PRISMGRID. A meta grid is a DUNE grid wrapper for any other DUNE grid, the so-called host grid. From the host grid's elements prismatic grid cells are constructed, leading to a higher-dimensional grid with a distinguished direction. In order to exploit the prismatic grid structure, several additional element iterators are available in DUNE-PRISMGRID. We compare the performance of the module against other meta grid implementations, and we show some numerical examples with DUNE-PRISMGRID.

1 Introduction

The DUNE-GRID module is the most mature part of the DUNE (Distributed and Unified Numerics Environment) project [9]. Within the framework described in [1, 2] a wide range of grids is covered including parallel and locally adaptive grids, geometrically non-conforming grids, and grids with curved linear elements.

At the time of this writing, the following grid implementations are contained in the DUNE-GRID module (for a detailed description we refer to [9]):

- the one-dimensional, adaptive grid `OneDGrid`,
- the structured grids `SGrid` and `YaspGrid` available for all grid dimensions,
- the grid wrappers for external grid managers `AlbertaGrid`, `ALUConformGrid`, `ALUCubeGrid`, `ALUSimplexGrid` and `UGGrid`,
- and the DUNE-GRID wrapper `GeometryGrid`, which wraps the geometrical representation of the underlying grid's elements.

Christoph Gersbacher
Section of Applied Mathematics, University of Freiburg,
Hermann-Herder-Str. 10, D-79104 Freiburg, Germany
e-mail: `christoph.gersbacher@mathematik.uni-freiburg.de`

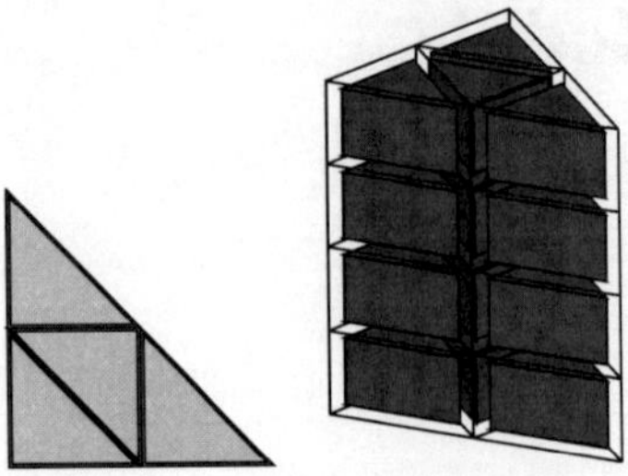

Fig. 1a Two-dimensional simplex host grid and resulting three-dimensional pentahedral `PrismGrid` with four layers.

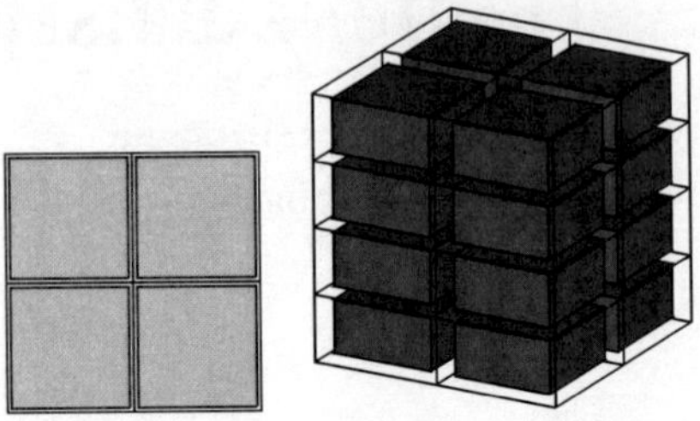

Fig. 1b Two-dimensional quadrilateral host grid and resulting three-dimensional hexahedral `PrismGrid` with four layers.

Implementations of the grid interface like `GeometryGrid` that are statically parametrized with another DUNE grid are called meta grids. The parameter grid within a meta grid is usually called the host grid.

In this paper, we describe the DUNE meta grid module DUNE-PRISMGRID. From every (leaf level) host grid element a number of prismatic cells are constructed in a conforming way and returned as a new DUNE grid (see Figures 1a, 1b). The geometrical representation of the prismatic cells relies strongly on the support of generic reference elements available in DUNE-GRID. Thereby, the implementation is fully generic, i.e., the module will work for all DUNE grid implementations and corresponding grid and world dimensions. The resulting grid features entities of all codimensions even if the host grid did not support subentities of higher codimensions. Additional element iterators are available extending the usual DUNE grid interface: column-wise, layer-wise, from lower to upper elements, and vice versa.

This paper is organized as follows: In Section 2 we give a short introduction to the DUNE-GRID framework. We describe the construction of prismatic cells and the prismatic grid. In Section 3 we give details on the usage and implementation of DUNE-PRISMGRID and an overview of additional features. In Section 4 we compare the performance of the implementation with other DUNE grid and meta grid implementations and show a more complex application using DUNE-PRISMGRID.

Throughout the paper, text in typewriter font (`PrismGrid`, ...) denotes actual class or method names.

2 General Framework

In this section, we give a short introduction to the DUNE-GRID framework as established in [1, 2]. Grids contain entities, and a reference element is associated with each grid element. The set of reference elements in the prismatic meta grid described here is different from that of the host grid. We therefore outline the importance of the generic reference elements and their implementation in DUNE-GRID for the design of DUNE-PRISMGRID. The description of prismatic grid cells and the prismatic grid is then straight-forward.

2.1 DUNE-GRID *Basics*

Following [1], a grid is a container of entities $\mathscr{E} = \{\mathscr{E}^0, \mathscr{E}^1, \dots, \mathscr{E}^d\}$ (for the sake of simplicity we do not consider a hierarchical grid). Here, d is the grid dimension, and $\mathscr{E}^c$ denotes the set of entities of codimension c. The corners, edges, faces, etc. of an entity are called subentities. For each entity $e \in \mathscr{E}$ there exists a reference element $R(e)$ such that the reference mapping

$$m_e : R(e) \to e \subset \mathbb{R}^w$$

is a homeomorphism that is consistent with the nested subentity relations on e (see [1, Definition 8, *Geometric realization*]). The number $w \geq d$ is called the world dimension.

The DUNE-GRID class `Grid` provides various iterators for accessing entities, and index and id sets for attaching data to these entities. The class `Entity<c>` represents an entity $e \in \mathscr{E}^c$, and stores the type of the reference element $R(e)$. In addition, each object of class `Entity` returns an object of type `Geometry`, which represents the geometric realization m_e.

2.2 *Generic Reference Elements*

Usually, a meta grid entity wraps a host grid entity. When asked for the reference element of the meta grid entity, in most meta grid implementations, it is sufficient to return the reference element of the host grid entity. However, this is not the case for a `PrismGrid`. As described below, a prismatic cell is of higher dimension than the underlying host grid entity. In addition, the reference elements are in general not unique for entities of higher codimensions. Instead, the correct type of the reference element has to be determined dynamically. The following concept of generic reference elements is therefore crucial for the fully generic implementation of `PrismGrid`.

By $\mathscr{R}^d$ we denote the set of d-dimensional generic reference elements in $\mathbb{R}^d$. It is characterized by the following assumptions:

- The only reference element in $\mathscr{R}^0$ is the point.
- From any reference element $R \in \mathscr{R}^d$ two higher-dimensional generic reference elements can be constructed: the *generic prism over* R, denoted by $R^{|} \in \mathscr{R}^{d+1}$, and the *generic pyramid over* R, denoted by $R^{\circ} \in \mathscr{R}^{d+1}$ (see Figures 2a, 2b).
- For each $R_{d+1} \in \mathscr{R}^{d+1}$, there is a generic reference element $R_d \in \mathscr{R}^d$ such that $R_{d+1} = R_d^{|}$, or $R_{d+1} = R_d^{\circ}$.

In DUNE-GRID, the `GenericReferenceElements` class provides the reference elements for all grid dimensions. An object of type `GenericReference-Element` can be created via a grid entity or a so-called topology id. Note that an

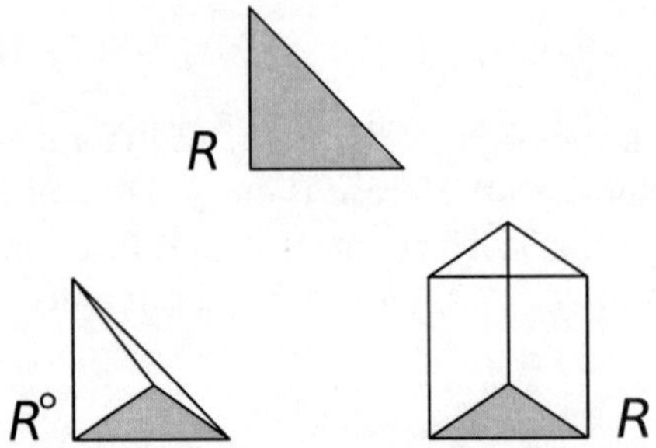

Fig. 2a Two-dimensional simplex reference element R, and generic reference elements R° and $R^{|}$ constructed from R.

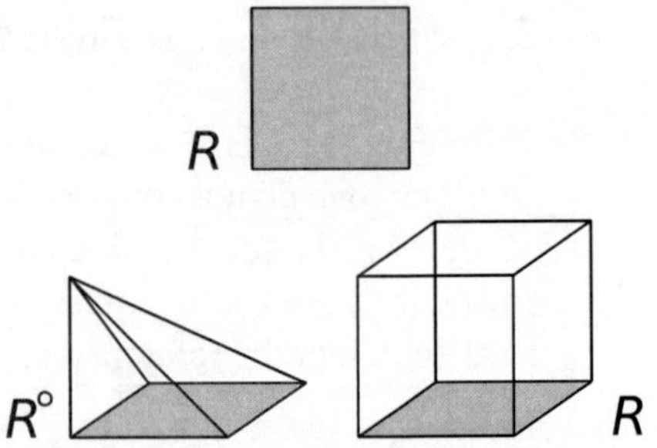

Fig. 2b Two-dimensional quadrilateral reference element R, and generic reference elements R° and $R^{|}$ constructed from R.

object of type `GenericReferenceElement` stores the types of the reference elements for all its corners, edges, faces, etc., as well.

For a more detailed description we refer to [8] and the article of A. Dedner and M. Nolte in this book.

2.3 Prismatic Grid

Let $\mathscr{H}$ be a d-dimensional grid in a w-dimensional world, called the host grid. The set of entities of codimension 0 in $\mathscr{H}$ will be denoted by $\mathscr{E}^0_{\mathscr{H}}$. Now, let L be a fixed number of layers and $\alpha_0 < \alpha_1 < \ldots < \alpha_L$.

By $p_{e,i}$ we will denote the prismatic cell constructed from an entity $e \in \mathscr{E}^0_{\mathscr{H}}$ and the interval $[\alpha_i, \alpha_{i+1}]$ for $i = 0, 1, \ldots, L-1$. The domain occupied by the prismatic cell is given by

$$\mathrm{dom}(p_{e,i}) = e \times [\alpha_i, \alpha_{i+1}] = \{ (x, x_{d+1}) \mid x \in e,\ x_{d+1} \in [\alpha_{i-1}, \alpha_i] \} \subset \mathbb{R}^{w+1}.$$

The subentities in $p_{e,i}$ are chosen in such a way, that the reference element $R(p_{e,i})$ corresponding to the prismatic cell is

$$R(p_{e,i}) = R(e)^{|},$$

where $R(e)$ denotes the reference element for e. By definition, top and bottom surface of the convex polytope $p_{e,i}$ are “flat”. Subentities parallel to the the newly added spatial direction are called “non-flat”.

A prismatic grid $\mathscr{P}(\mathscr{H})$ over $\mathscr{H}$ consists of all prisms constructed from $\mathscr{E}^0_{\mathscr{H}}$ and $(\alpha_i)_{i=0}^{L}$, i.e.,

$$\mathscr{E}^0_{\mathscr{P}(\mathscr{H})} = \{ p_{e,i} \mid e \in \mathscr{E}^0(\mathscr{H}),\ i = 0, 1, \ldots, L-1 \}.$$

The grid dimension of $\mathscr{P}(\mathscr{H})$ is $d+1$, and the world dimension is $w+1$. In general, a prismatic grid will by hybrid. If $\mathscr{H}$ is a conforming grid, $\mathscr{P}(\mathscr{H})$ will be conforming as well. In the newly added direction, grid cells will always align. At the time of this writing, the distribution of prismatic cells in this direction is restricted to the structured case $\alpha_i = \alpha_0 + i\Delta$, $\Delta = (\alpha_L - \alpha_0)/L$.

3 Using DUNE-PRISMGRID

In this section we give a short introduction on how to use DUNE-PRISMGRID. We show how an object of type `PrismGrid` is created, how the additional element iterators are used, and how to access the host grid wrapped within a `PrismGrid`. Note that for the time being a `PrismGrid` is not hierarchical.

3.1 PrismGrid Construction

Let `HostGrid` be a C++ type implementing the DUNE grid interface and `hostGrid` be an object of type `HostGrid`. The following code illustrates how to create an object of type `PrismGrid< HostGrid >`:

```
// number of layers
const int L = 4;
// interval boundaries
HostGrid::ctype alpha_0 = 0, alpha_L = 1;

// create grid
typedef PrismGrid< HostGrid > Grid;
Grid::LineGrid lineGrid( L, alpha_0, alpha_L );
Grid grid( hostGrid, lineGrid );
```

The internal type `LineGrid` is a very simple implementation of a structured one-dimensional grid for the interval `[alpha_0, alpha_L]`. The resulting grid will consist of `L` equally distributed layers (see Figures 1a, 1b).

In addition, the DUNE grid file (DGF) format for grid creation is supported by DUNE-PRISMGRID as well. A `PrismGrid` can be created either from a DGF interval block, or from a DGF file of the following structure best suited for more complex host macro grids:

```
DGF

HOSTGRID
macrogrids/hostgrid.dgf          % host grid filename
#

LINEGRID
0. 1. 4             % interval is [ 0., 1. ], 4 cells
#
```

Fig. 3a Iteration over a single column from lower to upper elements.

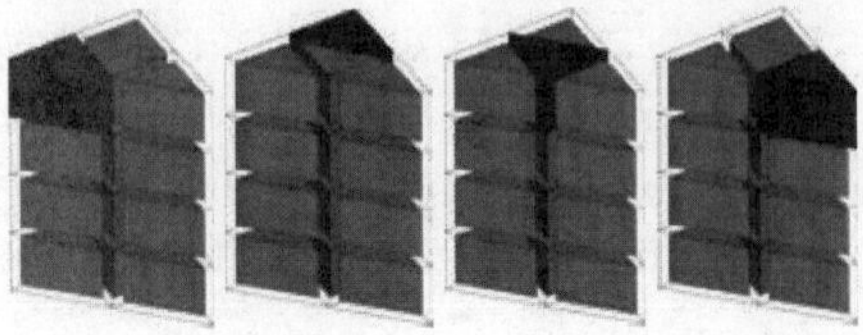

Fig. 3b Iteration over a single layer.

3.2 *Iterators*

An element iterator in DUNE-PRISMGRID is a combination of the host grid leaf iterator and a `LineGrid` iterator. There are several possible ways to iterate over the prismatic elements (see Figure 3a, 3b): column-wise, layer-wise, from lower to upper elements, and vice versa.

The standard choice for the leaf iterator returned by the grid is the column-wise traversal of elements. Thereby, internally only one full traversal of the host grid is needed. However, the default setting may easily be changed by the user. All iterators described above can be created independently in the following manner:

```
// types for layer iterator
typedef LayerIterator< Grid > LayerIterator;
typedef LayerIterator::Iterator Iterator;

// iterate from lower to upper layers
prismgrid::UpwindDirection direction = prismgrid::up;

// get begin and end iterator
Iterator it  = LayerIterator::begin( grid, direction );
Iterator end = LayerIterator::end( grid, direction );
```

3.3 *Determining the Reference Element*

As described in Section 2, the reference element of an entity has to be determined dynamically. Let `hostEntity` denote a host grid entity of dimension `dim`. The method `type()` returns the `GeometryType` of an entity. Since the introduction of the generic reference elements this class has not to be extended any more for grids introducing new reference element types in DUNE-GRID.

We wish to determine the reference element corresponding to the prismatic cell over `hostEntity`. The following piece of code shows how this is done in `PrismGrid`:

```
// get topologyId of host entity reference element
// and apply construction of generic prism
const unsigned int topologyId
  = hostEntity.type().id() | (1 << dim);
```

```
// get reference prism
typedef GenericReferenceElement< ctype, dim+1 >
  ReferencePrism;
const ReferencePrism & referencePrism
  = GenericReferenceElements< ctype, dim+1 >::
      general ( GeometryType( topologyId, dim+1 ) );
```

It is sometimes necessary to check whether a given `PrismGrid` entity of codimension `codim` > 0 is "flat", i.e., whether it lies on the upper or lower boundary of a codimension 0 `PrismGrid` entity. The class `Geometry` for example is specialized for these two situations. A subentity is identified by its `subIndex` relative to the entity it is contained in. The following strategy makes use of the enumeration of subentities in the generic reference elements:

```
// get host reference element
const GenericReferenceElement< ctype, dim > &
hostRfElem = GenericReferenceElements< ctype, dim >::
                general( hostEntity.type() );

// check whether subentity is flat for 0 < codim < dim+1
bool isFlat = ( hostRfElem.size(codim) <= subIndex );
```

3.4 Grid Refinement

For the time being, a `PrismGrid` is not hierarchical. Only leaf level entities, index set, iterators, etc., are available. However, the method `globalRefine` for global grid refinement is implemented. A single refinement results in calling the corresponding method on the host grid (possibly several times), while the number of intervals of the underlying `LineGrid` is doubled. Note that the grid width will be decreased by the factor 2:

```
// refine mesh globally by one refCount levels
void globalRefine ( int refCount )
{
  ...
  static int hostRefineStepsForHalf =
    Dune::DGFGridInfo< HostGrid >::refineStepsForHalf();
  for(int i = 0; i < refCount; ++i)
  {
    // call global refine on host grid
    hostGrid_->globalRefine( hostRefineStepsForHalf );
    // refine line grid
    lineGrid_->globalRefine( 1 );
  }
  ...
}
```

3.5 Access to Host Grid and Host Grid Entities

In many meta grid applications the meta grid and the host grid are used simultaneously. The `HostGridAccess` structure gives access to the underlying host grid stored in the meta grid:

```
const HostGrid & hostGrid =
  HostGridAccess< Grid >::hostGrid( grid );
```

The underlying host grid entity in a `PrismGrid` entity can be accessed in a similar way:

```
const HostGrid::Codim< 0 >::Entity & hostEntity =
  HostGridAccess< Grid >::hostEntity( entity );
```

4 Numerical Results

In this section, we compare the performance of the implementation of DUNE-PRISMGRID with other meta grid implementations in a standard Finite-Volume method in three spatial dimensions. Finally, we present a more complex example application using DUNE-PRISMGRID from the field of computational fluid dynamics.

4.1 Performance in an Explicit Finite-Volume Scheme

We investigate the performance of DUNE-PRISMGRID in an explicit Finite-Volume scheme. Let $\Omega = (0,1)^3$ and $a \in \mathbb{R}^3$. For given initial conditions c_0 and inflow boundary values b we want to compute the solution c of the linear transport problem

$$\left.\begin{aligned} \partial_t c + a \cdot \nabla c &= 0 && \text{in } \Omega \times]0, t_{end}[, \\ c &= b && \text{on } \Gamma_{in}, \\ c(\cdot, 0) &= c_0 && \text{in } \Omega. \end{aligned}\right\} \tag{1}$$

Here, $\Gamma_{in} = \{x \in \partial\Omega \mid a \cdot \nu(x) < 0\}$ is the inflow boundary, and $\nu(x)$ denotes the outer normal to $\partial\Omega$ in $x \in \partial\Omega$. On $\partial\Omega \backslash \Gamma_{in}$ no boundary condition is prescribed. The solution of (1) is computed using an explicit Finite-Volume scheme (for a description of the method we refer to [7]). For the test simulation, we chose $a = (1,1,1)$, $b = 0$, $t_{end} = 0.5$, and

$$c_0(x) = \begin{cases} 1, & \text{if } \frac{1}{8} < \|x\| < \frac{1}{2}, \\ 0, & \text{otherwise.} \end{cases}$$

We compare the performance of a number of hexahedral grids in three test runs with 96, 128, and 160 grid cells in each direction respectively:

- `ALUCubeGrid<3,3>`,
- `PrismGrid< ALUCubeGrid<2,2> >`,
- `IdentityGrid< ALUCubeGrid<3,3> >`,
- `IdGrid< ALUCubeGrid<3,3> >`.

`ALUCubeGrid<3,3>` is an unstructured hexahedral grid which serves as reference. For `PrismGrid` we use a two-dimensional `ALUCubeGrid` as host grid. In order to illustrate the possible performance loss possible with any meta grid, we take into account the identical grid wrappers `IdentityGrid` and `IdGrid`. `IdentityGrid` is a meta grid where all methods are called directly on the host grid. The resulting grid behaves exactly like the host grid. `IdGrid` is a second implementation of such an identical grid wrapper developed with regard to optimal performance by Martin Nolte. For more information on `IdGrid` see the article of R. Klöfkorn and M. Nolte in this book.

The computations were performed on an Intel Core i3 CPU with 3.07GHz. Figure 4 shows the computational time consumed for the whole numerical simulation of (1). The two identical grid wrappers illustrate the range of meta grid overhead possible (`IdGrid`: $< 15\%$, `IdentityGrid` $\approx 80\%$). With `PrismGrid< ALUCubeGrid<2,2> >`, the simulation takes around 53% to 62% more time compared to `ALUCubeGrid`. Being a generic meta grid wrapper, most methods are evaluated in the most general (and therefore most expensive) way. On the other hand, `PrismGrid` can make use of the structured grid nature in the newly added, vertical direction. In addition, the underlying `ALUCubeGrid<2,2>` contains significantly less elements than the hexahedral grid.

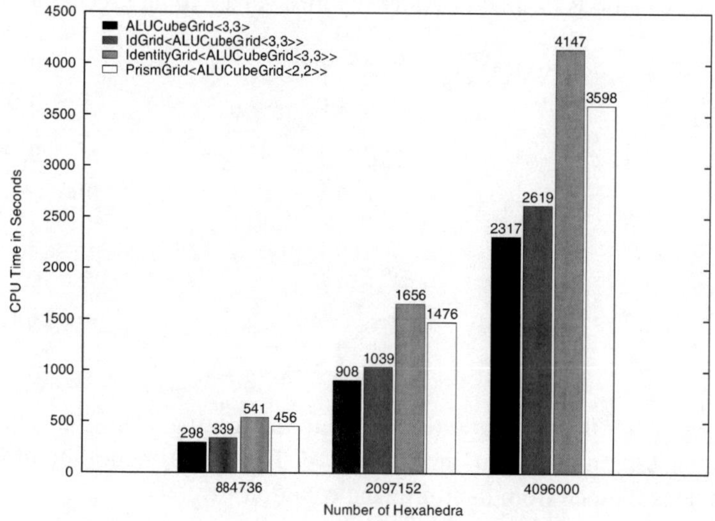

Fig. 4 Computational time consumed for the numerical solution of (1) with an explicit Finite-Volume scheme.

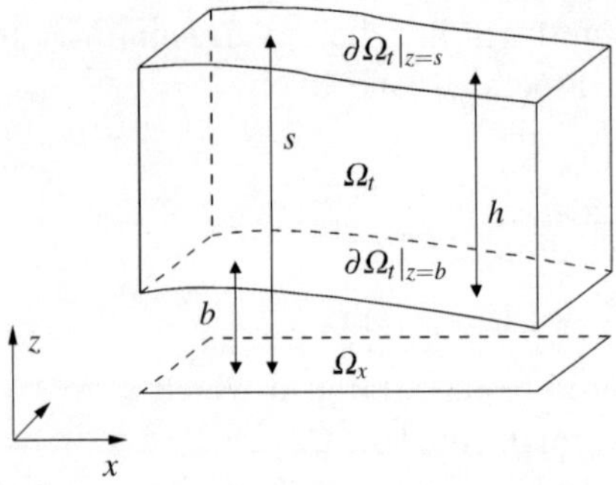

Fig. 5a The computational domain for the hydrostatic system.

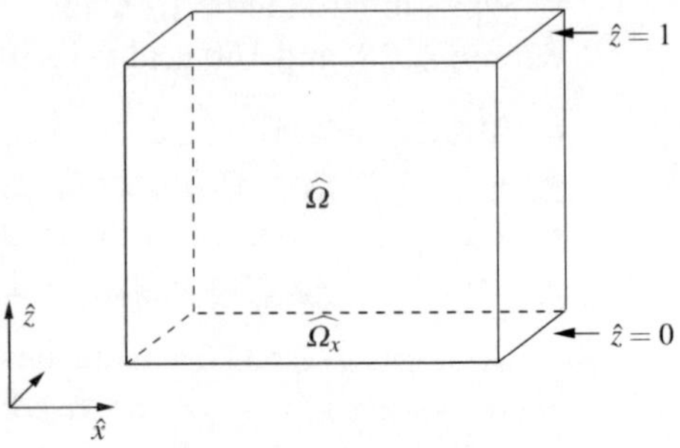

Fig. 5b The computational domain after σ-transformation.

4.2 *Application to Free-Surface Flows*

In this example we are interested in the numerical simulation of incompressible hydrostatic flows with a free surface. The domain occupied by the fluid at time t will be denoted by $\Omega_t \subset \mathbb{R}^d$, $d = 2,3$. By $u = (u_x, u_z)$ we will denote the d-dimensional velocity field, where u_z is the vertical velocity component.

We assume that for all times $t \in [0, t_{end}]$ it holds

$$\Omega_t = \left\{ (x,z) \in \mathbb{R}^{d-1} \times \mathbb{R} \mid x \in \Omega_x,\ b(x) < z < s(x,t) \right\},$$

where $b = b(x)$ and $s = s(x,t)$ are smooth functions representing the bottom topography and the position of the free surface respectively (see Figure 5a). By $h = s - b$ we denote the water height, and we assume that $h > 0$.

The following hydrostatic system was derived from the incompressible Navier-Stokes equations in shallow water regimes via a scaling process as presented in [5]:

$$\begin{aligned} \partial_t u_x + (u\cdot\nabla)u_x + \nabla_x p &= \mu \partial_z^2 u_x && \text{in } \Omega_t, \\ \nabla\cdot u &= 0 && \text{in } \Omega_t, \\ \partial_z p &= -g && \text{in } \Omega_t, \\ \partial_t s + u_x\cdot\nabla_x s &= u_z && \text{on } \partial\Omega_t|_{z=s} \end{aligned}$$

$$\begin{aligned} u_x\cdot\nabla_x b &= u_z && \text{on } \partial\Omega_t|_{z=b}, \\ \mu\partial_z u_x &= 0 && \text{on } \partial\Omega_t|_{z=s}, \\ \mu\partial_z u_x &= \kappa u_x && \text{on } \partial\Omega_t|_{z=b}, \end{aligned}$$

where $\kappa \geq 0$ is interpreted as bottom friction. On the lateral boundaries no-penetration conditions are prescribed. The evolution in time of the water height can be computed from the following equation

Fig. 6a 3d representation of initial data.

Fig. 6b Numerical solution to a later time.

$$\partial_t h + \nabla_x \cdot \Big(\int_b^s u_x \, dz \Big) = 0 \qquad \text{in } \Omega_x. \tag{2}$$

In [3, 6] a Local Discontinuous Galerkin discretization for this system is presented. In order to overcome the difficulty of the time-dependent domain the so-called σ-coordinate transformation

$$\hat{t} = t, \quad \hat{x} = x, \quad \hat{z} = \sigma(x,z,t) = \frac{z - b(x)}{h(x,t)}.$$

is applied to the system. Then, the transformed computational domain is fixed in time,

$$\hat{\Omega} = \{(\hat{x},\hat{z}) \mid (x,z) \in \Omega_t\} = \Omega_x \times (0,1), \tag{3}$$

see Figure 5b. The geometry of Ω_x may be very complex and may be captured by unstructured grids only. However, using an unstructured grid in all spatial dimensions would make the evaluation of of depth-averaged quantities as in (2) a non-trivial task. With DUNE-PRISMGRID, the geometry of the transformed fluid region (3) can be captured in a satisfactory way. For the computation of the integrals in (2) we can make use of the additional iterators described in Section 3.2.

The method was implemented using the discretization module DUNE-FEM [4, 10]. In Figures 6a, 6b the results of an example computation are shown in original physical quantities. We remark that for the visualization a smoothing operator was applied to the discrete height function. The numerical solution was computed using piecewise linear basis functions. Initial data is shown in the left picture in 6a, the velocity field is zero. Figure 6b shows the computed solution at a later time (for a full description of the test and the parameters chosen see [6]).

5 Conclusion

In this paper we presented the DUNE meta grid module DUNE-PRISMGRID. Higher-dimensional prismatic elements are constructed from an arbitrary DUNE host grid. The implementation is fully generic and strongly based on the generic reference elements in DUNE-GRID. We gave details on the implementation of several classes in DUNE-PRISMGRID and described additional features like the different element iterators. We examined the performance of the implementation in a Finite-Volume example, and presented an example using DUNE-PRISMGRID in the field of computational fluid dynamics. DUNE-PRISMGRID is still being developed. For the near future, we plan to add adaptivity and parallel support for the grid.

Acknowledgements. Contributions to DUNE-PRISMGRID have been provided by Martin Nolte and Andreas Dedner. The author would also like to thank the referees for their thorough reading and helpful remarks.

References

1. Bastian, P., Blatt, M., Dedner, A., Engwer, C., Klöfkorn, R., Ohlberger, M., Sander, O.: A Generic Grid Interface for Parallel and Adaptive Scientific Computing. Part I: Abstract Framework. Computing 82(2-3), 103–119 (2008)
2. Bastian, P., Blatt, M., Dedner, A., Engwer, C., Kornhuber, R., Klöfkorn, R., Ohlberger, M., Sander, O.: A Generic Grid Interface for Parallel and Adaptive Scientific Computing. Part II: Implementation and Tests in DUNE. Computing 82(2-3), 121–138 (2008)
3. Dedner, A., Gersbacher, C.: A Local Discontinuous Galerkin Discretization for Hydrostatic Free Surface Flows Using σ- transformed Coordinates. In: Proc. of the 13th International Conference on Hyperbolic Problems: Theory, Numerics, Applications, Beijing (2010) accepted for publication
4. Dedner, A., Klöfkorn, R., Nolte, M., Ohlberger, M.: A generic interface for parallel and adaptive discretization schemes: Abstraction principles and the DUNE-FEM module. Computing 90(3-4), 165–196 (2010)
5. Gerbeau, J.-F., Perthame, B.: Derivation of viscous Saint-Venant system for laminar shallow water; numerical validation. Discrete Contin. Dyn. Syst., Ser. B1 1(1), 89–102 (2001)
6. Gersbacher, C.: Local Discontinuous Galerkin Verfahren zur Simulation flacher dreidimensionaler Strömungen mit freier Oberfläche. Diploma thesis. University of Freiburg (2008)
7. Kröner, D.: Numerical Schemes for Conservation Laws. Wiley-Teubner, Chichester (1997)
8. Nolte, M.: Efficient Numerical Approximation of the Effective Hamiltonian. Doctoral dissertation. University of Freiburg (2011)
9. DUNE - Distributed and Unified Numerics Environment, `http://www.dune-project.org/`
10. The DUNE-FEM Module, `http://dune.mathematik.uni-freiburg.de/`

Performance Pitfalls in the DUNE Grid Interface

Robert Klöfkorn and Martin Nolte

Abstract. We discuss performance issues and common pitfalls one can encounter when dealing with the DUNE-GRID interface. This discussion includes the implementation of Cartesian grids in DUNE as well as the implementation of meta grids. Furthermore, for the use of local grid adaptivity we present several approaches to data restriction and prolongation and discuss their advantages and disadvantages in terms of performance. Finally, we also compare a general DUNE implementation of a Finite Volume scheme with a special purpose implementation.

1 Introduction

In this paper, we discuss performance issues related to the DUNE-GRID interface. A central design goal of DUNE-GRID is efficient support for Cartesian and unstructured grids through the same interface. We provide numerical results implying that, while making use of the Cartesian structure of a grid is a non-trivial task, an efficient implementation of such a grid is possible within the DUNE-GRID interface.

An abstract interface also has to enforce some general structure on its implementations, that could cause numerical algorithms to perform significantly slower than special purpose implementations. On the other hand, a meta grid, i.e., a DUNE grid implementing the interface through another DUNE grid, should ideally cause no extra loss in performance. We will give some hints on the efficient implementation of a meta grid and present some performance measurements on the overhead caused by a general meta grid.

Another important feature of the DUNE-GRID interface is local grid adaptivity. We will discuss different strategies for handling data restriction and prolongation within the DUNE context and their impact on the computational cost in a real

Robert Klöfkorn
Institut für Angewandte Analysis und Numerische Simulation
University of Stuttgart Pfaffenwaldring 57 D-70569 Stuttgart
e-mail: `robertk@mathematik.uni-stuttgart.de`

Martin Nolte
Section of Applied Mathematics, University of Freiburg
e-mail: `nolte@mathematik.uni-freiburg.de`

application. For this application, we also compare performance of a DUNE-based implementation with a special purpose implementation.

The paper is organized as follows: In Section 2 we mention some correct uses of the DUNE-GRID interface that might significantly decrease performance. Section 3 deals with the performance of different DUNE grid implementations on a Cartesian grid. We then address the performance of meta grids in Section 4. The performance of adaptive numerical simulations is discussed in Section 5. In this context, we also compare the DUNE-based implementation to a special purpose implementation. Finally, we draw some conclusions in Section 6.

2 Common Pitfalls

In this section we highlight common performance pitfalls when programming numerical schemes using the DUNE-GRID interface. We briefly discuss these on two examples, the use of entity pointers and the global id set.

An entity pointer essentially behaves like a smart pointer, i.e., it may delete the entity it points to when the pointer goes out of scope. Therefore, the following code may cause a segmentation fault:

```
const Entity &entity = *intersection.outside();
std::cout << entity.geometry().center() << std::endl;
```

The method `intersection.outside()` returns an `EntityPointer` object that might be destructed between the two lines (see [6, Section 12.2]), resulting in a dangling reference to the `Entity`. This issue can be resolved by using

```
const EntityPointer pEntity = intersection.outside();
const Entity &entity = *pEntity;
std::cout << entity.geometry().center() << std::endl;
```

Most grid implementations create the entities on demand, i.e., the entity only exists as long as at least one entity pointer points to it. In many cases, the entity implementation is even contained within the implementation of the entity pointer. In that case, copying the entity pointer means copying the entire entity. Notice that an entity might be a large object. For example, `AlbertaGrid` must keep a list of the father entities for each entity. Therefore, only a minimal number of entity pointers should be kept around.

In the DUNE-GRID interface, many objects, like the `Geometry` or the `Entity`, are returned by reference. Some grid implementations use this fact to create those objects on demand, i.e., at the time the user requests them. This generates additional overhead each time the object is requested, the worst case being that the object is filled on each request. Consequently, references to those objects should be stored for reuse. An example of ill-performant code is the following:

```
const Intersection &is = *intersectionIterator;
const int n = is.outside()->geometry().corners();
for( int i = 0; i < n; ++i )
  doSomething( is.outside()->geometry().corner( i ) );
```

The code can easily be rewritten to store references that are used more than once:

```
const Intersection &is = *intersectionIterator;
const EntityPointer ep = is.outside();
const Geometry &geo = ep->geometry();
const int n = geo.corners();
for( int i = 0; i < n; ++i ) doSomething( geo.corner( i ) );
```

Table 1 Comparison between the CPU time τ_1 for the first code snippet the CPU time τ_2 for the second code snippet for different 2-dimensional grid implementations.

Grid	τ_1	τ_2	τ_1/τ_2
AlbertaGrid	$182.8\,s$	$95.1\,s$	$192.1\,\%$
ALUCubeGrid	$77.6\,s$	$39.4\,s$	$197.1\,\%$
ALUSimplexGrid	$77.6\,s$	$50.5\,s$	$153.7\,\%$
UGGrid	$138.4\,s$	$52.9\,s$	$261.8\,\%$
SGrid	$104.4\,s$	$26.7\,s$	$391.8\,\%$
SPGrid	$5.7\,s$	$3.0\,s$	$192.4\,\%$
YaspGrid	$35.0\,s$	$13.8\,s$	$254.0\,\%$

A performance comparison for the two code snippets is shown in Table 1. It displays timings for the CPU time τ_1 needed for the first code snippet and τ_2 for the second code snippet evaluated 100 times on each intersection of a grid with 640×640 quadrilaterals (or twice as many triangles for AlbertaGrid and ALUSimplexGrid). As we see, depending on the grid implementation, the first code snippet requires an extra 50 % up to almost 300 % of CPU time in comparison to the second one.

Finally, let us have a look at the parallel grid interface. DUNE provides two id sets, a local and a global one, to uniquely identify an entity (see [2]). Since these id sets are artificial DUNE structures, a grid implementation might have to store them as data associated with the grid's native entity representation. Due to its less restrictive nature, a local id might be more readily available to the grid implementation than the global one. For example, a local id could be derived from a pointer to the native entity representation. A global id is necessarily a more complicated object, since it needs to be unique over all processes. Moreover, global ids have to remain persistent during load balancing, possibly leading to extra communication. Notice that, since id sets are returned by reference, the grid implementation cannot determine when the id set is no longer required. Once created, the grid cannot free resources required to provide the id set anymore. Therefore, uses of the global id set should be avoided whenever possible.

3 Performance of Cartesian Grids

A fast implementation of a structured (Cartesian) grid through the DUNE-GRID interface is a non-trivial task. It has even been said that `SGrid` is slow because it implements the complete grid interface. In this section we discuss design considerations for a fast implementation of a Cartesian grid in DUNE.

3.1 Implementing a Fast, Structured Grid: `SPGrid`

In this section we briefly present the main features and leading design goals of a new, parallel, structured grid implementation, `SPGrid` (see also [9]). As with `SGrid` and `YaspGrid`, the amount of memory required by `SPGrid` does not depend on the number of elements. It depends on the number of grid levels, though. Table 2 compares the features of `SGrid`, `YaspGrid`, and `SPGrid`. Apart from implementing the DUNE-GRID interface completely, `SPGrid` also provides additional features. For example, a bisection strategy can be used for global refinement.

Table 2 Feature comparison between `SGrid`, `YaspGrid`, and `SPGrid` in d space dimensions.

	`SGrid`	`YaspGrid`	`SPGrid`
provides entities of codimension	$0,\ldots,d$	$0, d$	$0,\ldots,d$
can communicate on codimension	—	$0, d$	$0,\ldots,d$
supported domains	$\prod[a_i,b_i]$	$\prod[0,b_i]$	$\prod[a_i,b_i]$
supports periodicity	no	yes	yes
supports anisotropic global refinement	no	no	yes
supports serial grids in parallel programs	—	no	yes
coordinate type as template parameter	yes	no	yes

On-the-fly objects, like the `Entity`, the `Geometry`, or the `Intersection`, should be minimal in size. Since the `Entity` implementation is required to return a `Geometry` by reference, `SPGrid` actually stores all data for the entity within its `Geometry` implementation. Moreover, geometric values that depend only on the grid level, like the `jacobian`, the `jacobianInverseTransposed`, or the `integrationElement`, are precomputed for each grid level. This is different from the implementations of `SGrid` and `YaspGrid`.

In a similar manner, `SPGrid`'s local geometries like the `geometryInFather` returned by the entity or the `geometryInInside` returned by the intersection need only be stored once for each level. The last observation also holds true for many unstructured grid implementations like `ALUCubeGrid`, `ALUSimplexGrid`, and `AlbertaGrid`.

3.2 Comparing Performance of Different Grid Implementations

To test the performance of SPGrid, we apply the example of an explicit Finite Volume scheme found in the DUNE-GRID-HOWTO [1]. In this example, the following advection problem for a scalar concentration c on the unit square $\Omega = [0,1]^d$ is solved:

$$\begin{aligned} \partial_t c + v \cdot \nabla c &= 0 && \text{in } \Omega \times]0,T[, \\ c &= b && \text{on } \Gamma_{in}, \\ c(\cdot,0) &= c_0 && \text{in } \Omega \end{aligned} \tag{1}$$

with a given divergence free velocity field v, initial concentration c_0, and boundary data b on the inflow boundary $\Gamma_{in} = \{(x,t) \in \partial\Omega \times]0,T[\,|\, v(x,t) \cdot \nu(x) < 0\}$. The exact choice of data influences the computation time only by the maximum time step, so we shall not detail them, here.

Explicit Finite Volume schemes for scalar advection equations are well suited for testing performance of grid implementations. They iterate over all grid elements and their intersections evaluating a very cheap numerical flux, so most of the CPU time is spent on the iteration and the evaluation of geometric quantities.

The CPU time required for the numerical solution of (1) up to $T = 0.5$ on different 3-dimensional (serial) grids is displayed in Table 3 and Fig. 1. The computation was performed on an Intel Xeon X5460 CPU with 3.16 GHz. As we can see, SPGrid is almost twice as fast as YaspGrid and more than 8 times as fast as SGrid, thus proving the efficiency of the SPGrid implementation.

Table 3 CPU time for the Finite Volume solution of (1) on different grid implementations.

Hexahedra	Time Step	SGrid	YaspGrid	SPGrid	ALUCubeGrid	UGGrid
884 736	0.003 437	295.6 s	61.8 s	33.4 s	281.3 s	3422.8 s
2 097 152	0.002 578	923.4 s	190.5 s	104.9 s	871.3 s	10548.3 s
4 096 000	0.002 063	2231.4 s	477.3 s	271.8 s	2192.8 s	26001.5 s
884 736	0.003 437	885.0%	185.0%	100.0%	842.2%	10247.9%
2 097 152	0.002 578	880.3%	181.6%	100.0%	830.6%	10055.6%
4 096 000	0.002 063	821.0%	175.6%	100.0%	806.8%	9566.4%

The performance of two fully unstructured hexahedral grid implementations, ALUCubeGrid and UGGrid, has also been measured. Since these grids do not know that our mesh is actually Cartesian, they have to perform a non-negligible overhead to dynamically compute constant data like normals or volumes. Moreover, since both implementations store each element in memory, there is a notable increase in the required memory bandwidth. What we see, though, is that ALUCubeGrid still performs better than SGrid while UGGrid is around 100

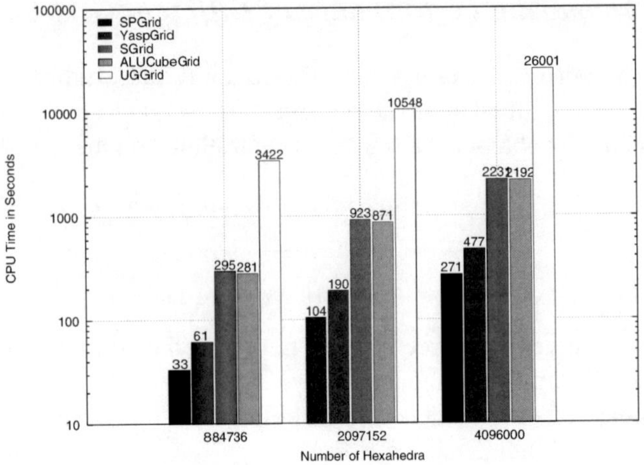

Fig. 1 CPU time for the Finite Volume solution of (1) on different grid implementations.

times slower than `SPGrid`. To some extent, this reflects the different design goals of `ALUCubeGrid` and `UGGrid`. While `ALUCubeGrid` was designed to be used with explicit Finite Volume schemes for systems of conservation laws, `UGGrid` was designed for solving partial differential equations via multigrid methods.

From this performance test, we can draw two conclusions. Though `SPGrid` implements a superset of the features of both, `SGrid` and `YaspGrid`, it is a lot faster than either of the two. This demonstrates that the efficient implementation of structured grids within the DUNE-GRID interface requires special care. Secondly, the comparison with the unstructured grids shows that the choice of the grid implementation should strongly depend on the problem to be solved. It demonstrates that despite the work overhead caused by the use abstract interfaces, like the DUNE-GRID interface, we gain the flexibility to simply choose a suitable grid implementation. The potential performance gain justifies the need for such abstract interfaces in scientific computing.

4 Performance Overhead of Meta Grids

Meta grids implement the DUNE-GRID interface based on another DUNE grid (called *host grid* in this context), usually adding some features. An example of a meta grid contained in DUNE-GRID is the `GeometryGrid`, which replaces the host grid's geometry implementation. Wrapping `YaspGrid` into a `GeometryGrid` we can, for example, obtain a parallel grid modeling a helix. Other examples of meta grids are `PrismGrid` and `MultiDomainGrid` (see the corresponding articles in this book).

In this section we consider the overhead necessarily added by a meta grid. To this end, we consider two metagrids, `IdGrid` and `IdentityGrid`, which do not add any functionality at all. Ideally, these grids should perform exactly as fast as their corresponding host grid.

4.1 Minimal Overhead for Meta Grids

The DUNE-GRID interface currently causes considerable overhead for a meta grid. As an example we consider the data section of the IdGrid entity implementation:

```
class IdGridEntity {
  ...
  const HostEntity *hostEntity_;
  mutable Geometry geo_;
};
```

Since entities cannot be copied, the meta grid's entity implementation must either hold a pointer to the host grid's entity (as done by IdGrid) and a corresponding entity pointer (as done by IdentityGrid). Additionally, it must provide memory for the geometry, since it is returned by reference. In the case of IdGrid, the geometry merely contains a pointer to the host grid's geometry. All in all, the IdGrid entity implementation adds two extra pointers of memory overhead.

Efficient wrapping of the entity pointer even requires type casting. The IdGrid entity pointer looks as follows:

```
class IdGridEntityPointer {
  ...
  operator const EntityPointerImp & () const  {
    return reinterpret_cast< const EntityPointerImp & >( *this );
  }
  mutable Entity entity_;
  HostIterator hostIterator_;
};
```

Apart from the Entity, which has to be returned by reference, it contains the host iterator, i.e., an instance of the host grid's class derived from EntityPointer. This complicated construct is caused by the required derivation of all iterators from the corresponding entity pointer implementation. The reinterpret_cast can can also be found in the EntityPointer interface class. Unfortunately, its result is explicitly specified as undefined in general (see [6, Section 5.2.10]).

An alternative approach is taken by IdentityGrid. The entity pointer implementation just stores a host grid's entity pointer; the iterator then stores the host grid's iterator. In this case, each meta grid entity pointer stores two entity pointers from the host grid, yielding a non-negligible performance loss.

Other classes, like the index sets, require similar overhead. Index sets have to store a pointer (or reference) to the host grid's index set. Each time an index is requested, an additional pointer might have to be dereferenced, depending on inlining.

4.2 *Comparing the Performance of* `IdGrid` *and* `IdentityGrid`

In this section we again measure the performance for the explicit Finite Volume example from the DUNE-GRID-HOWTO (see Section 3) for both meta grids with varying host grids.

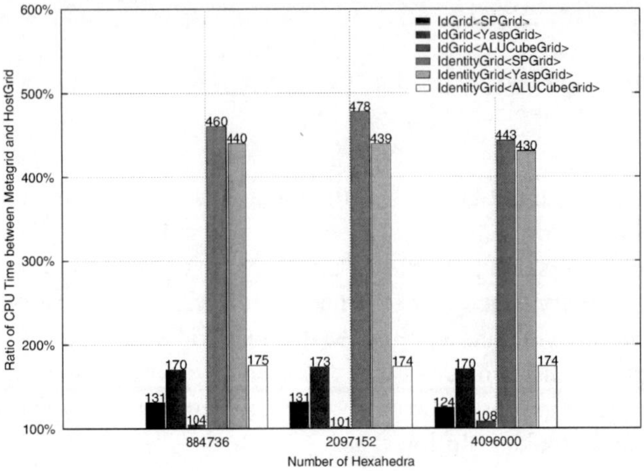

Fig. 2 Relative CPU time for the Finite Volume solution of (1) on `IdGrid` and `IdentityGrid` for different host grids; 100% marks the CPU time required on the host grid.

In Fig. 2 we present the relative performance overhead caused by `IdGrid` and `IdentityGrid` for the Finite Volume solution of (1). For `IdGrid`, the performance overhead caused by the meta grid ranges from approximately 5% in the case of `ALUCubeGrid` to around 70% in the case of `YaspGrid`. In the case of `IdentityGrid`, this overhead ranges from approximately 75% up to 350%. Notice that `ALUCubeGrid` requires around 4 times the CPU time of `SPGrid`. This partly explains the small overhead required for `ALUCubeGrid` in comparison to `SPGrid`.

Overall, writing meta grids using the DUNE-GRID interface is a good way to extend the list of features of several grids. Nevertheless, the implementation of the meta grid has to be done carefully to avoid dramatic performance decay. Also, some design issues in the DUNE-GRID interface, such as returning the `Geometry` by reference, still have to be resolved to decrease the performance loss caused by using meta grids.

5 Performance of Adaptive Simulations

In this section we study the performance of adaptive computations of the compressible Euler equations using an explicit Finite Volume scheme. A more detailed survey is given in [7].

5.1 *Comparison of Adaptive Grids in* DUNE

In this subsection we discuss the implementation of adaptivity using the DUNE-GRID interface. While the adaptation of the grid itself is already provided, the implementation of the data restriction and prolongation is left to the user. We show that there are several ways to do this, not all of them are suitable for an explicit Finite Volume scheme.

To study the performance of our adaptive computations we consider the Euler equations of gas dynamics in a domain $\Omega \subset \mathbb{R}^d$. In conservative variables $u = (\rho, \rho\mathbf{v}, \varepsilon)^T \in \mathbb{R} \times \mathbb{R}^d \times \mathbb{R}$, these equations read:

$$\partial_t \begin{pmatrix} \rho \\ \rho\mathbf{v} \\ \varepsilon \end{pmatrix} + \nabla \cdot \begin{pmatrix} \rho\mathbf{v} \\ \rho\mathbf{v}\mathbf{v}^T + p(u)\mathbb{I} \\ (\varepsilon + p(u))\mathbf{v} \end{pmatrix} = 0 \quad \text{in } \Omega \times]0, T[,$$

where $I \in \mathbb{R}^{d\times d}$ denotes the identity matrix. The equations are closed by the pressure law for an ideal gas: $p(u) = (\gamma - 1)(\varepsilon - \frac{1}{2}\rho|\mathbf{v}|^2)$, where $\gamma = 1.4$ denotes the adiabatic constant.

The discretization used is an explicit Finite Volume scheme with the HLLEM numerical flux [5]. For further reading we refer to [8].

The algorithm is divided into 5 sub-steps:

1. (*Communication*) copy data from `interior` to `ghost` cells (in parallel only),
2. (*Flux evaluation*) calculation of the numerical fluxes,
3. (*Synchronization*) calculate the minimal time step size,
4. (*Evolution*) update the solution to the next time level,
5. (*Adaptation and load balancing*) the grid is adapted and (in parallel runs) load balanced if necessary.

In the following the performance of the code is measured by calculating $\eta_E = \sum_i \eta_{i,E}$, the average run time per element for one timestep, where $\eta_{i,E}$ is the CPU time for sub-step i. The detailed description of the setting is found in [7].

In the following we analyze different restriction and prolongation strategies for adaptive algorithms using DUNE-GRID and DUNE-FEM [3]. These are:

fem-generic. In the generic restriction/prolongation strategy from DUNE-FEM restriction is done before the grid is modified, prolongation is done afterwards.

fem-callback. The callback strategy from DUNE-FEM restricts or prolongates data ad-hoc when an element is removed or created. This is realized by passing a callback object satisfying the `RestrictionProlongation` interface to the grid's `adapt` method. This method is only supported by `AlbertaGrid` and `ALUGrid`.

dune-level. This restriction and prolongation is proposed in the DUNE-GRID how-to [1]. Restriction and prolongation is done level-wise, using a `LevelIterator`, and storing all data attached to leaf elements into a `std::map`. Association of user data for the transfer from storages used in the calculation phase to the intermediate containers is done using the persistent index maps from DUNE [2, Definition 26]. This is also the main disadvantage of this method since the data has to be moved to the intermediate containers and back even though only a small amount of elements might be refined or coarsened.

In Table 4 we present the results calculated with `ALUSimplexGrid` for the above mentioned restriction and prolongation techniques in combinations with different data storages. These are:

LeafIndexSet. The data is stored in a `std::vector<double>` like array indexed by the the standard `LeafIndexSet` provided by a DUNE grid. During adaptation the data has to be transferred to a `std::map` using the `LocalIdSet` for identification.

generic `PersistentContainer`. The `PersistentContainer` is an implementation of a container that identifies data with grid entities. The generic implementation is based on a `std::map` using the `LocalIdSet` for identification.

specialized `PersistentContainer`. The `PersistentContainer` has been specialized for `ALUGrid` and `AlbertaGrid` to improve performance of data access. This implementation stores data in a `std::vector` like array.

DUNE-FEM also provides an implementation of a consecutive index set that can be used in adaptive simulations without data transfer to intermediate containers. In Table 4 this is represented by *Index in `PersistentContainer`*. For details we refer to [3, 7]. Results for the following test using `AlbertaGrid` or `UGGrid` can be found in [7].

We discover that storing the data in the specialized `PersistentContainer` and using the fem-callback strategy performs best when using `ALUSimplexGrid`. However, storing data in a `PersistentContainer` is not suitable for implicit schemes where linear solvers come into play, since data storage in `PersistentContainer`s is in general non-consecutive. For this reason, instead of storing data directly in the specialized `PersistentContainer` we only store a consecutive index set. The difference to the standard DUNE-GRID index sets is that this new index set can be used also in adaptive computations without unnecessary data transfer to intermediate containers. In Table 4 we see that using the index set stored in the specialized `PersistentContainer` does not lead to a serious performance loss in comparison to storing the data directly in the specialized `PersistentContainer`. However, in comparison to the standard DUNE techniques it is much faster for this example.

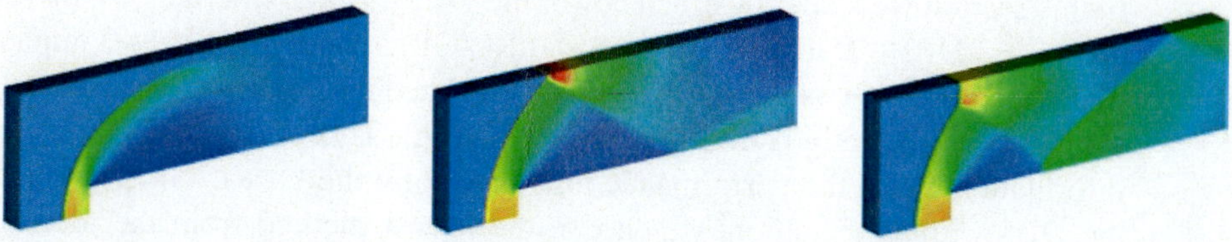

Fig. 3 Density distribution at times $T = 0.5, 1.5, 3.5$, from left to right.

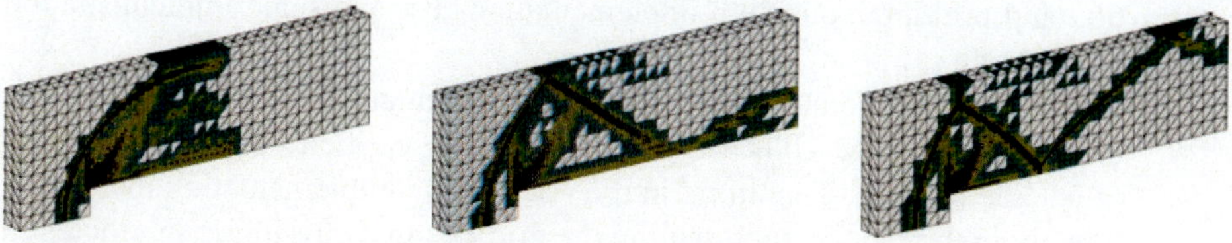

Fig. 4 Locally adapted grids at times $T = 0.5, 1.5, 3.5$, from left to right.

Table 4 Restriction and prolongation with `ALUSimplexGrid`. For easy comparison we define $\eta^* := 8.053 \cdot 10^{-4}$ and $\eta_5^* := 7.424 \cdot 10^{-5}$, which are the values from the fastest run (see page 53 for the definition of $\eta_{s,E}$ and η_E).

Method	$\eta_{2,E}$	$\eta_{4,E}$	$\eta_{5,E}$	η_E	$\eta_{5,E}/\eta_5^*$	η_E/η^*
			Data in Specialized `PersistentContainer`			
fem-callback	$7.280 \cdot 10^{-4}$	$3.081 \cdot 10^{-6}$	$7.424 \cdot 10^{-5}$	$8.053 \cdot 10^{-4}$	1.00	1.00
fem-generic	$7.304 \cdot 10^{-4}$	$3.014 \cdot 10^{-6}$	$1.554 \cdot 10^{-4}$	$8.888 \cdot 10^{-4}$	2.09	1.10
dune-level	$7.285 \cdot 10^{-4}$	$3.076 \cdot 10^{-6}$	$4.054 \cdot 10^{-4}$	$1.137 \cdot 10^{-3}$	5.46	1.41
			Index in Specialized `PersistentContainer`			
fem-callback	$7.686 \cdot 10^{-4}$	$2.632 \cdot 10^{-6}$	$1.018 \cdot 10^{-4}$	$8.730 \cdot 10^{-4}$	1.37	1.08
fem-generic	$7.770 \cdot 10^{-4}$	$2.851 \cdot 10^{-6}$	$1.972 \cdot 10^{-4}$	$9.771 \cdot 10^{-4}$	2.66	1.21
dune-level	$7.555 \cdot 10^{-4}$	$3.107 \cdot 10^{-6}$	$4.325 \cdot 10^{-4}$	$1.191 \cdot 10^{-3}$	5.83	1.48
			Index in Generic `PersistentContainer`			
fem-callback	$9.438 \cdot 10^{-4}$	$2.589 \cdot 10^{-6}$	$7.332 \cdot 10^{-4}$	$1.680 \cdot 10^{-3}$	9.88	2.09
fem-generic	$9.477 \cdot 10^{-4}$	$3.478 \cdot 10^{-6}$	$8.237 \cdot 10^{-4}$	$1.775 \cdot 10^{-3}$	11.10	2.20
dune-level	$8.910 \cdot 10^{-4}$	$2.916 \cdot 10^{-6}$	$6.897 \cdot 10^{-4}$	$1.584 \cdot 10^{-3}$	9.29	1.97
			`LeafIndexSet`			
dune-level	$7.186 \cdot 10^{-4}$	$2.650 \cdot 10^{-6}$	$5.437 \cdot 10^{-4}$	$1.265 \cdot 10^{-3}$	7.32	1.57

5.2 Efficiency of the Interface in Parallel Adaptive Simulations

In this section we study the efficiency of the DUNE-GRID interface. To this end, we compare a DUNE (DUNE-COMMON, DUNE-GRID, and DUNE-FEM) implementation of a Finite Volume scheme with an specialized implementation based on ALUGRID [4, 10] directly (DIRECT implementation). This comparison should give us information about the performance loss caused by the DUNE-GRID interface. For the DUNE implementation, we choose the fastest method from the previous section, i.e., storing the DoFs directly in the specialized `PersistentContainer` of `ALUSimplexGrid` (or `ALUCubeGrid`) and using fem-callback for restriction and prolongation. Both implementations use the same implementation of the HLLEM flux [5].

The main difference between the two implementations concerns the storage of data. Originally, ALUGRID was designed with explicit Finite Volume schemes in mind (see [4, 10]). Therefore, in the ALUGRID implementation the data is stored directly in the objects representing the grid elements, resulting in efficient data access and allocation. In this sense the DIRECT implementation is optimal for the chosen discretization scheme and can be considered a hard contestant for efficiency measurements. In DUNE on the other hand, data is stored separately from the grid, in this case a vector.

Since grid adaptation is performed in each time step, the execution time for the grid modification is comparable to the cost of the numerical scheme (about 10–20% of the overall time). In this sense the explicit Finite Volume scheme is a challenging problem for a general grid interface like DUNE where data is managed independently from the grid. Anyhow, the data storage in DUNE should lead to some advantage in the evolution step of the algorithm (sub-step 4, see enumeration on page 53) which is expected to be faster than in the DIRECT implementation.

Table 5 Performance loss for the serial DUNE code. $\theta_{s,E} := (\eta_{s,E}^{dune} - \eta_{s,E}^{direct})/\eta_E^{direct}$ and θ_E is calculated accordingly (see page 53 for the definition of $\eta_{s,E}$ and η_E).

element type	$\theta_{2,E}$	$\theta_{4,E}$	$\theta_{5,E}$	θ_E
tetrahedra	9.06 %	-5.02 %	3.2 %	7.24 %
hexahedra	15.80 %	-3.32 %	2.64 %	15.10 %

In Table 5 we present the performance loss of the code when using the DUNE grid interface. For the tetrahedral elements we loose around 7% and for the hexahedral elements the loss is 15%. The main performance loss comes from the flux calculation, where iterations over grid elements takes place. As in the previous sections we see that the design of the grid interface is not yet perfect in terms of performance. On the other hand, the comparison of general code with a special purpose code is a challenging example. Thus, a performance loss between 7% and 20% is more than acceptable, keeping in mind that this loss will decrease for more expensive schemes.

In Table 6 we present the comparison of speedup $s_{L\to K}$ from L to K, $L < K$, processors for both parallel implementations. The optimal speedup would be K/L. On the left the results for the direct code are shown; on the right we have the corresponding results for the DUNE code. The computations were performed on the XC4000 of the SCC Karlsruhe. Although the numerical problem is rather simple, both codes show very good *strong scaling*. This indicates that the DUNE-GRID interface is well capable of adaptive computations including dynamic load balancing.

Table 6 Speedup and efficiency measured with respect to the **average run time per element and timestep** for a run with $4 \to 32$ processors (tetrahedra) and $8 \to 64$ processors (hexahedra) using a fixed sized problem.

(a) DIRECT implementation (tetrahedra)

K	η_K	$s_{4\to K}$	$\approx s_{4\to K}/4$
4	1.7730e-03		
8	9.8920e-04	1.79	0.90
16	5.2784e-04	3.36	0.84
32	2.8455e-04	6.23	0.78

(b) DUNE implementation (tetrahedra)

K	η_K	$s_{4\to K}$	$\approx s_{4\to K}/4$
4	2.6427e-03		
8	1.2329e-03	2.14	1.07
16	6.5431e-04	4.04	1.01
32	3.4282e-04	7.71	0.96

(c) DIRECT implementation (hexahedra)

K	η_K	$s_{8\to K}$	$\approx s_{8\to K}/8$
8	1.4958e-03		
16	8.1890e-04	1.83	0.91
32	4.3945e-04	3.40	0.85
64	2.4875e-04	6.01	0.75

(d) DUNE implementation (hexahedra)

K	η_K	$s_{8\to K}$	$\approx s_{8\to K}/8$
8	1.8302e-03		
16	1.0149e-03	1.80	0.90
32	5.5265e-04	3.31	0.83
64	3.1091e-04	5.89	0.74

6 Conclusions

We presented some critical points in the DUNE-GRID interface in terms of performance of numerical codes. There are several pitfalls, some of which were mentioned. Numerical evidence to the feasibility of a fast implementation of a Cartesian grid, i.e., `SPGrid`, was demonstrated.

The use of meta grids helps to develop new features while reusing existing code. However, such implementations have to be designed carefully in order to avoid unnecessary losses in performance.

Local grid adaptivity is available in DUNE and, as shown, can be used efficiently in adaptive, parallel computations requiring dynamic load balancing.

In conclusion we have shown that, despite its abstract nature, the DUNE-GRID interface allows, with some care, the efficient implementation of demanding numerical codes.

References

1. Bastian, P., Blatt, M., Dedner, A., Engwer, C., Klöfkorn, R., Ohlberger, M., Sander, O.: The Distributed and Unified Numerics Environment DUNE-GRID interface how-to (2008),
`http://www.dune-project.org/doc/grid-howto/grid-howto.pdf`
2. Bastian, P., Blatt, M., Dedner, A., Engwer, C., Klöfkorn, R., Ohlberger, M., Sander, O.: A generic grid interface for parallel and adaptive scientific computing. part I: Abstract framework. Computing 82(2-3), 103–119 (2008)
3. Dedner, A., Klöfkorn, R., Nolte, M., Ohlberger, M.: A generic interface for parallel and adaptive scientific computing: Abstraction principles and the DUNE-FEM module. Computing 90(3-4), 165–196 (2010)
4. Dedner, A., Rohde, C., Schupp, B., Wesenberg, M.: A parallel, load balanced MHD code on locally adapted, unstructured grids in 3D. Comput. Vis. Sci. 7(2), 79–96 (2004)
5. Einfeldt, B., Munz, C., Roe, P., Sjögreen, B.: On Godunov-type methods near low densities. J. Comput. Phys. 92(2), 273–295 (1991)
6. Programming languages — C++. International Standard ISO/IEC 14882 (2003)
7. Klöfkorn, R.: Numerics for evolution equations — a general interface based design concept. Ph.D. thesis. Albert-Ludwigs-Universität Freiburg (2009),
`http://www.freidok.uni-freiburg.de/volltexte/7175/`
8. Kröner, D.: Numerical Schemes for Conservation Laws. Wiley-Teubner (1997)
9. Nolte, M.: The dune-spgrid module,
`http://dune.mathematik.uni-freiburg.de/grids/dune-spgrid/`
10. Schupp, B.: Entwicklung eines effizienten Verfahrens zur Simulation kompressibler Strömungen in 3D auf Parallelrechnern. Ph.D. thesis. Albert-Ludwigs-Universität Freiburg (1999),
`http://www.freidok.uni-freiburg.de/volltexte/68/`

DUNE-MULTIDOMAINGRID: A Metagrid Approach to Subdomain Modeling

Steffen Müthing and Peter Bastian

Abstract. We present a DUNE-GRID extension that enhances existing DUNE grids with the ability to designate arbitrary subsets of their leaf entity complex as subdomains and present them as new grid objects. We describe the functionality of this module, which is called DUNE-MULTIDOMAINGRID and available as free software, and outline its implementation. In particular, we highlight the performance characteristics of our module and present a way of tailoring them to a specific problem by means of a modular backend engine. Finally, we give some pointers to current applications of the module.

1 Introduction

Many interesting problems today involve the simultaneous examination of multiple physical models which interact by means of a set of coupling conditions. As many physical models are based on partial differential equations (PDEs), the resulting simulations will often be based on Finite Elements or similar methods. Additional challenges arise when those simulations also involve multiple spatial subdomains, e.g. investigations of multi-domain problems like fluid-structure interaction. Unfortunately, the majority of established simulation frameworks in this area, like deal.II [3], FEniCS [14] or DUNE [4, 5], are mostly focused on solving problems on a common domain. Moreover, the standard approach to modeling these subdomains in a simulation has been to employ a distinct grid for each subdomain, mostly

Steffen Müthing
Institut für Parallele und Verteilte Systeme, Universität Stuttgart
e-mail: steffen.muething@ipvs.uni-stuttgart.de

Peter Bastian
Interdisziplinäres Zentrum für Wissenschaftliches Rechnen, Universität Heidelberg
e-mail: peter.bastian@iwr.uni-heidelberg.de

motivated by the desire to reuse existing software packages for the subproblems. One major drawback of this approach is the great care required during the creation of the individual grids, especially in the case of interface problems with complex boundary geometries, where both sides of the interface need to match the common geometric shape of the interface as closely as possible. Moreover, any calculation of coupling conditions, whether on a lower-dimensional interface or on overlapping subdomains, requires a conforming matching of the underlying grid discretizations, which for distinct grids will often necessitate the calculation of a common subtessellation. One software package that can be used for this purpose in the context of the DUNE framework is the module DUNE-GRIDGLUE [6]. Examples of more comprehensive software packages that are capable of handling inter-program mesh and DoF transfer and coordinating a weakly coupled solution scheme include SIERRA [10], MpCCI [15].

In the following, we will present an extension module for the DUNE toolkit that takes a novel approach towards the problem of representing meshes for problems with multiple subdomains. This module, called DUNE-MULTIDOMAINGRID, allows the user to track multiple spatial subdomains within a single master grid and presents those subdomains as distinct grids to the user. We begin by describing the functionality and programming interface of the module. In Section 3, we outline the implementation of the module and demonstrate its favorable performance scaling characteristics. After pointing out some first applications in Section 4, we finish with a short conclusion and an outlook towards ongoing and possible future enhancements.

2 Using MultiDomainGrid

`MultiDomainGrid` has been developed as an add-on module for DUNE-GRID and can be found in the DUNE module DUNE-MULTIDOMAINGRID. This module can be obtained from a GIT repository at [17]. It is free software and available under the same license as the DUNE core modules (the GNU General Public License with a special runtime exception, for further details see [2]). In addition to DUNE-GRID and its dependencies, an installation of the Boost C++ libraries [1] is also required. In particular, the code uses the Boost packages MTL, Fusion and smart_ptr.

2.1 Meta Grids

One of the most important goals of the DUNE framework is flexibility, enabling users to easily exchange parts of their simulation, e.g. the grid manager. Thus, high modularity is a central design objective. For this purpose, the components involved in building a PDE solver are precisely defined on a mathematical level. Moreover, there is an exact specification of the programming interface of conforming implementations. The most important one of these interfaces is the DUNE grid interface, described in [4, 5]. This interface is sufficiently fine-grained to allow for an implementation of a new grid on top of an existing DUNE grid implementation, creating a so-called *meta grid*. Meta grids are a powerful mechanism for enhancing or modifying the functionality of an existing grid, instantly providing new functionality to a wide range of different grid managers without requiring an understanding of their implementations, as any delegation happens through the common grid interface.

Due to the versatility and easy implementability of the concept, meta grids have attracted considerable interest in recent times. Examples within DUNE include `SubGrid` [13], `GeometryGrid` (part of DUNE-GRID) and `PrismGrid` [12].

2.2 Overview

The functionality and basic design principle of DUNE-MULTIDOMAINGRID are shown in Figure 1. The module is implemented in terms of two cooperating meta grids, `MultiDomainGrid` and `SubDomainGrid`: The `MultiDomainGrid` wraps the existing *host grid*, provides an interface for setting up and accessing subdomains and contains all subdomain-related data structures, while the individual subdomains are exposed as `SubDomainGrid` instances, which are very lightweight objects that combine the information from the host grid and the associated `MultiDomainGrid` to present the subdomains as regular DUNE grids. In order to differentiate between the different subdomains, they are assigned numbers from the set $[0, N-1]$, where N denotes the maximum number of subdomains supported by the grid.

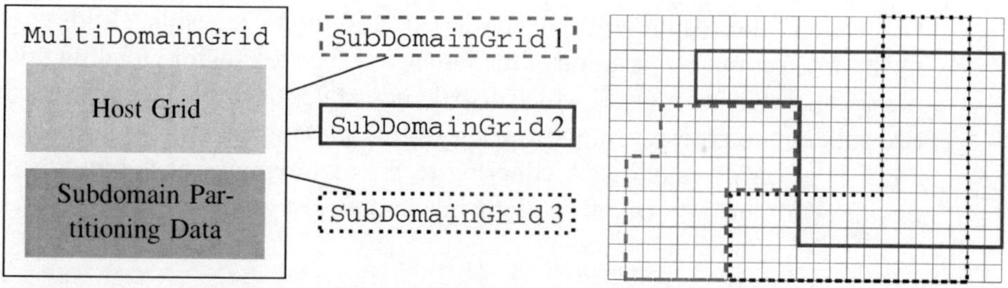

Fig. 1 Functionality and basic design of DUNE-MULTIDOMAINGRID: An existing host grid is wrapped and extended to support multiple subdomains, each available to the user as a distinct `SubDomainGrid`.

A `MultiDomainGrid` retains all capabilities of the underlying grid, including full support for adaptivity and parallelism if these are provided by the host grid.

2.3 Grid Creation

To add multi-domain functionality to an existing grid, it has to be wrapped in a `MultiDomainGrid`:

```
typedef MultiDomainGrid<HostGrid,FewSubDomainsTraits<
        HostGrid::dimension,maxSubDomains> > MDGrid;
bool supportLevelViews = true / false;
MDGrid mdgrid(hostgrid,supportLevelViews);
```

Here, the second template parameter is a policy class that can be used to customize internal data structures and algorithms for different application scenarios. Right now, there are two possible choices, `FewSubDomainsTraits` and `ArrayBasedTraits`, which are explained in detail in Section 3.2. In the case shown here, `maxSubDomains` denotes the maximum number of subdomains supported by the grid. It is possible to disable support for level views in the subdomains, which can substantially reduce the memory requirements of the grid.

2.4 Subdomain Setup

Subdomains are designated by an interface that closely resembles the existing interface for grid adaptation. All of the involved methods are defined on the `MultiDomainGrid`. Modifying the subdomain layout and adapting the grid are mutually exclusive processes; while one of the two is in progress, the other one cannot be started.

Every subdomain comprises all levels of the grid, and the extent of a subdomain is defined by marking *leaf grid* cells as belonging to the subdomain. Afterwards, this information is propagated up the grid hierarchy, but *not* back down again, which in fact violates the requirements of the DUNE grid interface, as the children of a cell must form a complete partition of the father cell. The reason for violating this principle is illustrated in Figure 2, which also demonstrates the basic mechanism, showing two subdomain grids resulting from a leaf grid marking pattern: As the coarsest grid consists of only a single cell, adhering to the requirements of the interface would not allow for any non-trivial subdomains on this host grid.

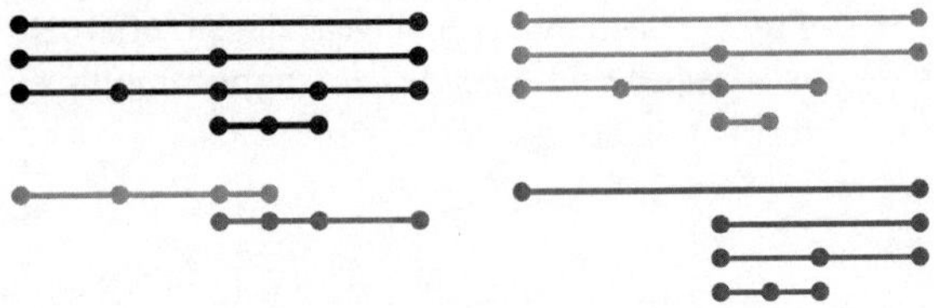

Fig. 2 Hierarchic subdomain construction from marked leaf grid: host grid hierarchy and marked leaf subdomains (left), resulting subdomain grid hierarchies (right)

The complete programming interface for the marking process is defined in terms of member methods of the `MultiDomainGrid`: The process must be started by calling `beginSubDomainMarking()`. This sets up the necessary data structures to simultaneously manage the old and the new subdomain layout. Afterwards, the methods

```
addToSubDomain(SubDomainType s, const Entity& e);
removeFromSubDomain(SubDomainType s, const Entity& e);
assignToSubDomain(SubDomainType s, const Entity& e);
removeFromAllSubDomains(const Entity& e);
```

can be used to modify the current subdomain layout by calling them with codimension 0 entities of the leaf grid. Once the new layout has been completely built up, `preUpdateSubDomains()` propagates the subdomain membership information to all codimensions and to the level grids. It also rebuilds the index maps based on the new subdomain layout. After this step, the user should do any required data projection between the old and the new subdomain state. A subsequent call to `updateSubDomains()` activates the new subdomain layout, but retains all information about the old state, so that further user data transfer can occur. This information can finally be deleted by calling `postUpdateSubDomains()`.

2.5 Subdomain Usage

As already mentioned, subdomains are accessible as DUNE grid objects and support the complete interface with the only exception of grid-morphing operations like grid adaptation or load balancing. Note that `SubDomainGrid`s are – like all grids – noncopyable and can thus only be obtained by reference by calling

```
MultiDomainGrid::SubDomainGrid&
MultiDomainGrid::subDomain(SubDomainType s);
```

2.5.1 Inter-grid Data Transfer

While any calculations that are restricted to a single subdomain can be performed using the corresponding `SubDomainGrid` in a completely transparent fashion, it is often necessary to convert grid objects like entities between the different grids they are contained in, i.e. the host grid, the `MultiDomainGrid` and the `SubDomainGrid`. For this purpose, the two meta grids provide the following methods:

```
template<...>
class MultiDomainGrid {

  const HostEntity&
  hostEntity(const MultiDomainEntity& e);
```

```
  HostEntityPointer
  hostEntityPointer(const MultiDomainEntity& e);

  const MultiDomainEntity&
  multiDomainEntity(const SubDomainEntity& e);

  MultiDomainEntityPointer
  multiDomainEntityPointer(const SubDomainEntity& e);

};

template<...>
SubDomainGrid {

  SubDomainEntityPointer
  subDomainEntityPointer(const MultiDomainEntity& e);

};
```

The last method will fail if passed an entity that does not belong to the subdomain. There is a corresponding set of methods for `Intersections`, which for brevity is not shown here.

In addition, the `MultiDomainGrid` can also be queried for the set of subdomains a given entity belongs to. This information can be retrieved with the method

```
const MDGridTraits::Codim<0>::SubDomainSet&
IndexSet::subDomains(const MultiDomainEntity& e);
```

`SubDomainSet` is an implementation-defined class that offers iterator-based access to the subdomain indices, subdomain membership testing and information about the set size. Note that the `SubDomainSet` implementation may vary between codimensions for efficiency reasons.

2.6 Subdomain Interface Extraction

Many algorithms require the traversal of the interface between two subdomains. We represent the elements of such an interface using the class `SubDomainInterface`, which is similar to a standard DUNE `Intersection`, but offers two additional methods

```
SubDomainIndexType subDomainInInside();
SubDomainIndexType subDomainInOutside();
```

for retrieving the subdomains of the two adjacent entities. Two iteration scenarios stand out as particularly useful and are directly supported by custom iterators in `MultiDomainGrid`:

Visiting the interface between two specific subdomains. Given two subdomains s_1 and s_2, this iterator will visit all intersections between entities e_i and e_j where e_i is contained in s_1 but not in s_2 and e_j is contained in s_2 but not in s_1. These iterators can be obtained using the methods

```
LeafSubDomainInterfaceIterator
leafSubDomainInterfaceBegin(SubDomainIndexType sd1,
                            SubDomainIndexType sd2);

LeafSubDomainInterfaceIterator
leafSubDomainInterfaceEnd(SubDomainIndexType sd1,
                          SubDomainIndexType sd2);
```

Note that the equivalent methods for level view iterators were omitted for brevity.

Visiting all subdomain interfaces in the grid. If an application needs to iterate over several subdomain interfaces, the iterators described above are not very efficient, because every iteration entails a full traversal of the underlying host grid. On the other hand, we can efficiently calculate the set of subdomain interfaces a given grid intersection belongs to: Given the intersection between two grid cells e_1 and e_2 belonging to the sets of subdomains S_1 and S_2, respectively, let $D_1 = S_1 \setminus S_2$ and $D_2 = S_2 \setminus S_1$. Then the set of all subdomain intersections is given by the tensor product $D_1 \times D_2$. As outlined in Section 3.2, the per-entity set of subdomains is usually stored as a bitset in an integral type, making these set operations very efficient.

An efficient iteration scheme based on this algorithm that visits all subdomain interfaces of a `MultiDomainGrid` is implemented in the iterators obtained from the methods

```
LeafAllSubDomainInterfacesIterator
leafAllSubDomainInterfacesBegin();

LeafAllSubDomainInterfacesIterator
leafAllSubDomainInterfacesEnd();
```

and their level view equivalents. These iterators visit all subdomain-subdomain interfaces contained in the `MultiDomainGrid` and require only a single host grid traversal. It is thus often very beneficial to phrase application-level algorithms in such a way that all subdomain-subdomain related tasks can be handled in parallel using these iterators, yielding a performance benefit especially for larger numbers of subdomains that are small in comparison to the complete domain.

3 Implementation

One major problem related to meta grids in DUNE is the overhead associated with the wrapper layer around the underlying grid: While the DUNE grid interface makes heavy use of C++ template programming and is designed in a way that most of it can be optimized away by a good compiler, it was not designed with the possibility of creating meta grids in mind. Consequently, a certain amount of information duplication has to take place in the wrapper, making it expensive to "stack" several meta grids on top of each other.

3.1 Design

The design and implementation of DUNE-MULTIDOMAINGRID was inspired by the module DUNE-SUBGRID [13], but it significantly expands on the concepts demonstrated there, as that module was built for a different purpose and is only capable of tracking a single subdomain in an efficient manner. As shown by Gräser and Sander, most of the required functionality of the grid can be reused unchanged from the host grid. Since individual subdomains simply represent a subset of the entities contained in the host grid, the only functionality that needs to be changed for the subdomain tracking are those parts of the grid interface concerned with the entity complex of the host grid. This includes entity and intersection iterators, entity counts and index maps, but not id maps, as the ids generated by the host grid are still unique and can be reused. Implementing any of these features requires a subdomain set map $\sigma : E \to \mathscr{S}$ that associates every entity $e \in E$ with the set of subdomains $s \subseteq S$ it is contained in. Here, S is the set of all subdomains and $\mathscr{S} := \mathscr{P}(S)$ denotes the set of all possible subdomain combinations. As our new functionality has to be built on top of the DUNE grid interface, σ has to be stored as user data using one of the mechanisms offered by the interface, i.e. either entity indices and arrays or entity ids and sets. Since σ will be evaluated very often during operations on the subdomains, we opted for the former approach, as the $\mathscr{O}(\log N)$ complexity of C++ sets would have had a major performance impact.

Like Gräser and Sander, we assume that the majority of the spatial domain will be covered by at least one subdomain and conclude that all additional data required for supporting subdomain grids is most easily stored within the `IndexSet` class of the DUNE grid interface. While it is possible to devise a number of different storage schemes for this data, the one implemented by `MultiDomainGrid` is depicted in Figure 3. The two containers for codimension and geometry type are due to the fact that the index maps of the grid interface map are defined in terms of the geometry type, i.e. an index map I_{gt} for entities of geometry type gt is a map $E_{gt} \to [0, |E_{gt}| - 1]$. It is thus necessary to create a distinct data structure for each geometry type contained in the grid. This data structure is an array of tuples (s,i), where s denotes the set of subdomains the entity belongs to, and i represents an index.

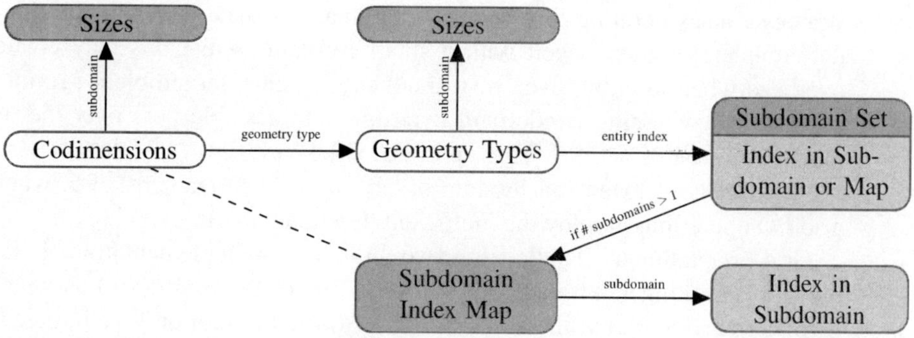

Fig. 3 Basic storage scheme for subdomain data in `MultiDomainGrid`

The precise meaning of i depends on the cardinality of s: if $|s| = 1$, it represents the index of the entity within the single subdomain it belongs to. Otherwise, it refers to an entry in a second array which contains a map $\lambda_e : s \to I$.

3.2 *Storage Backends*

The algorithm behind the storage strategy described above is currently fixed, but it can nevertheless be influenced by choosing different implementations for several map and set containers. In particular, it is possible to specify the types of the containers shaded in darker grey in Figure 3. Right now, there are two pre-defined policies that can be picked when creating a `MultiDomainGrid`:

- `FewSubDomainsTraits`: This policy is optimized for regular multi-physics problems. It allows for up to 64 subdomains that may overlap in an arbitrary fashion and maintains $\mathcal{O}(1)$ complexity for testing subdomain membership of an entity and for looking up a subdomain entity index. The limit on the number of subdomains is due to the fact that the subdomain set is stored as a bitmask in an integral type.
- `ArrayBasedTraits`: An alternative policy designed for large numbers of subdomains. It does not place any intrinsic limit on the maximum number of subdomains, but limits the number of subdomains a single entity can belong to. Both subdomain membership testing and subdomain index lookup require a single binary search of a sorted array spanning the local set of subdomains and are thus of complexity $\mathcal{O}(\log|s_e|)$.

3.3 *Efficiency*

Entity and intersection iterators are implemented on top of the host iterators, skipping entities not contained in the subdomain and modifying returned intersections if an adjacent cell does not belong to the subdomain. This implementation strategy causes iteration time over a subdomain to scale linearly with the size of the complete domain, albeit with a small constant. While this may create some overhead when iterating over very small subdomains, the problem can mostly be avoided by coalescing subdomain iterations into a single pass over the underlying `MultiDomainGrid`. This approach is taken in DUNE-MULTIDOMAIN, our DUNE-PDELAB extension module based on `MultiDomainGrid`, where only grid I/O operations employ the inefficient iteration pattern.

Grid adaptation is handled transparently by the implementation. It is possible to place refinement marks on both the `MultiDomainGrid` as well as any `SubDomainGrid`, but the actual grid transformation can only be initiated by the `MultiDomainGrid`, a choice that was made to explicitly emphasize the fact that in general, refinement performed on one subdomain will also change the structure of other subdomains, because all subdomains share the common discretization provided by the host grid.

In order to evaluate the runtime and memory overhead of `MultiDomainGrid`, we took a simple example program from DUNE-PDELAB that solves the Poisson equation in 2D on the unit square using a mix of Dirichlet and Neumann boundaries. This program was modified to run either directly on the host grid, on the `MultiDomainGrid` or on a `SubDomainGrid` spanning the complete domain. This way, all three program variants solve the exact same problem in an identical fashion, which allows for a good assessment of the overhead imposed by wrapping the host grid and by using a grid defined on a subdomain, respectively. We ran the benchmark using both a structured (`YaspGrid`) and an unstructured (`ALUSimplexGrid`) host grid to investigate whether our module exhibits a different behavior on those two types of grids. The grids for the benchmark were generated by starting with a single square (or two triangles in the case of `ALUSimplexGrid`) covering the unit square and iteratively refining those macro grids. All results were obtained by running the simulations 10 times and averaging the numbers obtained from the individual runs.

host grid	`YaspGrid`				`ALUSimplexGrid`			
grid size	262144				524288			
Operation	t_{host} [s]	$\frac{t_{MD}}{t_{host}}$	$\frac{t_{SD}}{t_{host}}$	$\frac{t_{SD}}{t_{MD}}$	t_{host} [s]	$\frac{t_{MD}}{t_{host}}$	$\frac{t_{SD}}{t_{host}}$	$\frac{t_{SD}}{t_{MD}}$
GFS setup	0.177	1.19	2.36	1.99	0.720	1.55	3.18	2.05
constraints evaluation	0.313	1.46	4.66	3.19	0.821	1.62	4.09	2.53
pattern construction	1.14	1.04	1.21	1.16	1.38	1.16	1.64	1.42
residual evaluation	1.00	1.02	1.55	1.52	2.04	1.15	1.79	1.56
jacobian evaluation	3.56	1.03	1.09	1.05	5.11	1.05	1.17	1.11

Fig. 4 Performance comparison between `MultiDomainGrid` (t_{MD}), associated `SubDomainGrid` (t_{SD}) and underlying host grid (t_{host}) for common DUNE-PDELAB operations. Grids were refined 9 times.

We assessed the runtime overhead of our grid by timing several standard DUNE-PDELAB operations which all involve a grid iteration, but vary in the computational effort per grid cell, ranging from the very fast grid function space setup to the evaluation of the Jacobian by numerical differentiation. The results of the comparison can be found in Figure 4. Those results clearly show a considerable performance overhead which might be reduced by further optimization of the wrapper implementation. In particular, the `SubDomainGrid`-based variant exhibits a disproportionate runtime increase, which is probably linked both to the fact that it is implemented as a meta grid on top of the `MultiDomainGrid`, but also to the additional subdomain membership checks required during the iteration. In general, the performance penalty is more pronounced for simple and fast operations, making the grid in its current state more suited to numerical schemes that involve a moderate to large computational effort per cell.

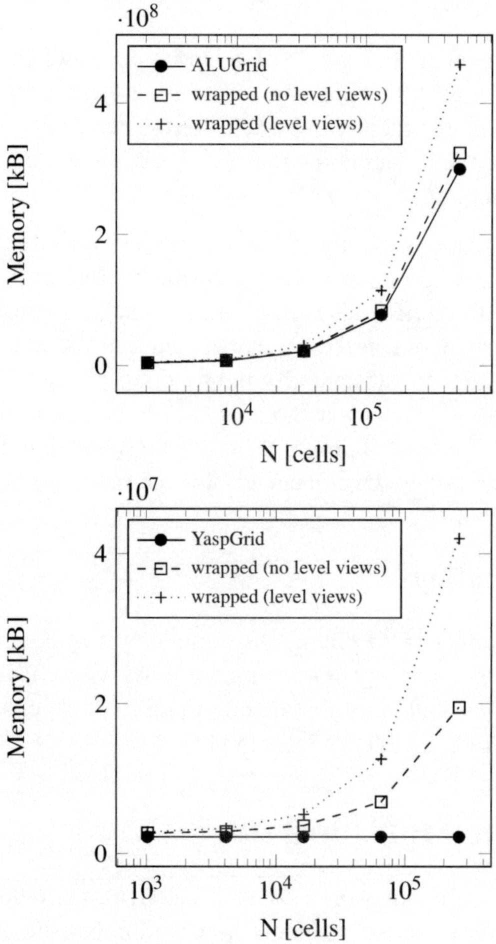

Fig. 5 Memory usage of `MultiDomainGrid` and underlying host grid for `ALUGrid` (unstructured) and `YaspGrid` (structured).

The additional memory requirements of `MultiDomainGrid` are illustrated in Figure 5. The memory usage of the programs was measured directly after grid creation (for the host grid) and after subdomain assignment (for `MultiDomainGrid`). The results show that while the design of our grid necessarily changes the storage characteristics of a structured grid, losing the constant size property, the scaling behavior of an unstructured grid remains unchanged. Unfortunately, the total amount of extra memory required is still rather substantial. In order to mitigate this problem, it is possible to reduce the storage requirements of a `MultiDomainGrid` in two ways:

Remove support for unused codimensions. When developing a simulation using Finite Volumes or a Discontinuous Galerkin scheme, only entities of codimension 0 are required. In this case, it is possible to selectively deactivate entities with unused codimensions in the `SubDomainGrids`. Note that these disabled entities are only removed from the `SubDomainGrids`, they remain accessible for code directly working on the `MultiDomainGrid`.

Deactivate level grid views. Non-adaptive codes will normally only access the leaf grid view of a grid. If a program never accesses the level grid views of the subdomains, support for them can be removed. The results of this optimization can be seen in Figure 5, reducing the memory overhead for unstructured grids to a mostly negligible amount.

These optimizations mainly affect the memory requirements of the module and do not affect the runtime performance during normal grid operations. They do, however, reduce the time it takes to rebuild the subdomain information after changing the subdomain layout, as there are fewer entities that need to be tracked and assigned indices. In order to substantially improve the runtime efficiency of the module, it would probably be necessary to adopt several changes to the DUNE grid interface. There have been several proposals in this direction, but discussions of these proposals among the DUNE developers are still ongoing.

4 Applications

The first application of this module employs it in the context of a multi-scale approach in the domain of porous media flow, where it is used to designate subdomains of a fine-scale grid that belong to individual cells of the related coarse-scale grid. For further details, we refer to [11].

4.1 DUNE-MULTIDOMAIN

In the context of our work, DUNE-MULTIDOMAINGRID will not be used on its own, but only as one component in a comprehensive framework for the solution of multi-domain multi-physics problems called DUNE-MULTIDOMAIN [16]. This framework is built on DUNE-PDELAB [7] and provides extensions for defining

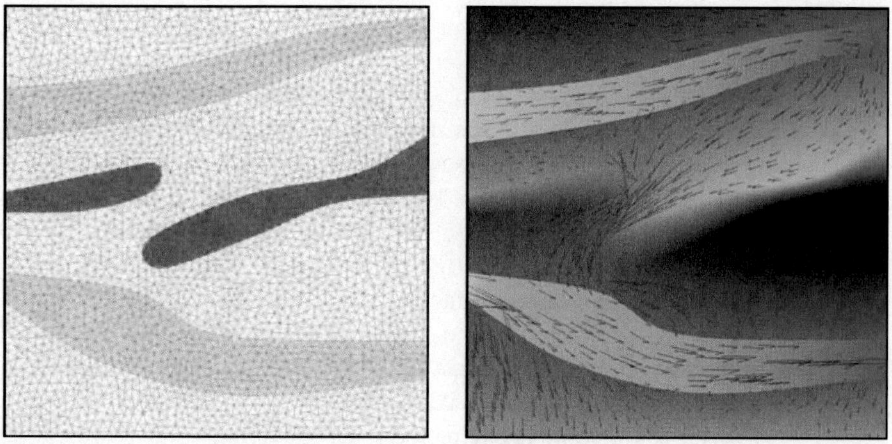

Fig. 6 Advanced example built using the DUNE-PDELAB extension module DUNE-MULTIDOMAIN. Left: Computational grid with two free-flow channels from left two right and two impermeable areas in the porous medium. Right: Pressure and velocity fields of an example simulation.

variables and residual forms on different subdomains, while retaining the capability to assemble and solve a resulting multi-domain problem in a strongly coupled fashion. This module is currently still in development, although some first results using a strongly coupled model of a porous medium and a free flow domain using standard Beavers-Joseph conditions (based on the model presented in [8]) have already been derived. Due to our flexible subdomain handling, these simulations can easily be applied to complex geometries and advanced models which employ different discretizations on different parts of the domain.

An example of the results obtained in this context is shown in Figure 6, which shows a fully coupled simulation based on a mixed formulation with a standard Taylor-Hood element in the Stokes domain and a primal formulation using a Discontinuous Galerkin scheme of third order for the porous medium. Note the advantage of using a DG approach in the porous medium, which greatly improves the ability of the model to capture the jump in permeability of 10^{-5} between the bulk of the porous medium and the two low permeability regions extending from the left and the right border of the domain.

As we are mostly interested in investigating the problem assembly process, the resulting linear problem was solved using the direct solver library SuperLU [9].

Delegating the subdomain management to DUNE-MULTIDOMAINGRID provides several advantages in this context:

- The definition of the individual subdomains is completely decoupled from the actual problem solver, which only expects a `MultiDomainGrid` with subdomains set up to a certain scheme (e.g. Stokes = 0, Darcy = 1). This way, it is easy to change the method used for subdomain designation (analytical definition, physical regions in loaded mesh file, ...).

- Defining variables on the subset of a grid view is difficult in DUNE, as for efficient access all user data has to be stored in arrays based on the mapping provided by an `IndexSet`, a storage scheme that is ill-suited to data that is only defined on parts of a grid view. `MultiDomainGrid` solves this problem in a general way by allowing the user to designate subdomains of a grid and presenting those as `SubDomainGrids`, on which variables can be defined normally. This way, we are able to reuse the elaborate function space mechanism from DUNE-PDELAB in DUNE-MULTIDOMAIN with minimal changes.
- The standard solution output facilities of DUNE based on VTK always write a complete grid view. In order to easily output variables which are only defined on parts of the domain, we can again leverage the fact that a subdomain defined on a `MultiDomainGrid` is exposed as a DUNE grid and use a standard `VTKWriter` for variables defined on that subdomain.
- We intend to extend DUNE-MULTIDOMAIN to handle moving subdomains. This requires a general mechanism for detecting cells which are added to and removed from a subdomain, as these cells will in general require special handling, similar to the restriction / prolongation steps in adaptive simulations. This information can easily be extracted from `MultiDomainGrid` (cf. Section 2.4).

Handling all these tasks at the grid level has allowed us to keep the implementation of DUNE-MULTIDOMAIN much simpler than would have been possible otherwise and to focus on creating a framework that is sufficiently abstract to maybe also incorporate different subdomain management backends at a later stage. One possibility in this context would be a backend based on DUNE-GRIDGLUE [6].

5 Conclusions and Outlook

We have presented a novel approach to the problem of mesh creation and mesh traversal for multi-domain problems based on the idea of enhancing existing grids in a way that allows them to be divided up into a number of subdomains. These subdomains appear as fully-functional grids in their own right, facilitating the re-use of existing solver implementations for subproblems defined on those subdomains. Our implementation supports a wide variety of subdomain layout scenarios, including arbitrary overlaps and modification of the subdomain layout over the course of a simulation. By constructing our software as a meta grid on top of the DUNE grid interface, the new functionality immediately becomes available to all existing DUNE grids without requiring any internal modifications.

One important area of improvement is the issue of load-balancing behavior. Currently, the module reuses the load-balancing algorithm of the host grid unchanged, but this will in general be suboptimal, as different subdomains may impose different computational effort during simulation. For this purpose, we intend to investigate the possibility of assigning weights to the different subdomains and use these to influence the size of the partitions generated by the load-balancing algorithm of the host grid.

Finally, while the functionality of DUNE-MULTIDOMAINGRID may greatly simplify the development of multi-domain simulations, our ultimate goal is the creation of a multi-domain simulation framework operating at a higher abstraction level. For this purpose, an extension of DUNE-PDELAB is currently in development and has already been used to produce some initial results on the problem of Stokes-Darcy coupling. As outlined in Section 4, the development of such a high-level multidomain discretization framework is greatly simplified by delegating the subdomain management to DUNE-MULTIDOMAINGRID.

References

1. The Boost C++ libraries, `http://www.boost.org`
2. Dune download and license page, `http://www.dune-project.org/download.html`
3. Bangerth, W., Hartmann, R., Kanschat, G.: deal.II Differential Equations Analysis Library, Technical Reference, `http://www.dealii.org`
4. Bastian, P., Blatt, M., Dedner, A., Engwer, C., Klöfkorn, R., Kornhuber, R., Ohlberger, M., Sander, O.: A Generic Grid Interface for Parallel and Adaptive Scientific Computing. Part I: Abstract Framework. Computing 82(2-3), 103–119 (2008)
5. Bastian, P., Blatt, M., Dedner, A., Engwer, C., Klöfkorn, R., Kornhuber, R., Ohlberger, M., Sander, O.: A Generic Grid Interface for Parallel and Adaptive Scientific Computing. Part II: Implementation and Tests in DUNE. Computing 82(2-3), 121–138 (2008)
6. Bastian, P., Buse, G., Sander, O.: Infrastructure for the coupling of dune grids. In: Proceedings of ENUMATH 2009, pp. 107–114 (2010)
7. Bastian, P., Heimann, F., Marnach, S.: Generic implementation of finite element methods in the Distributed and Unified Numerics Environment (DUNE). Kybernetika 46(2), 294–315 (2010)
8. Cao, Y., Gunzburger, M., Hu, X., Hua, F., Wang, X., Zhao, W.: Finite element approximations for stokes-darcy flow with beavers-joseph interface conditions. SIAM Journal on Numerical Analysis 47(6), 4239–4256 (2010)
9. Demmel, J.W., Eisenstat, S.C., Gilbert, J.R., Li, X.S., Liu, J.W.H.: A supernodal approach to sparse partial pivoting. SIAM Journal on Matrix Analysis and Applications 20(3), 720–755 (1999)
10. Edwards, H.: Managing complexity in massively parallel, adaptive, multiphysics applications. Engineering with Computers 22, 135–155 (2006)
11. Flemisch, B., Darcis, M., Erbertseder, K., Faigle, B., Lauser, A., Mosthaf, K., Müthing, S., Nuske, P., Tatomir, A., Wolff, M., Helmig, R.: DuMux: DUNE for Multi-Phase, Component, Scale, Physics, . . . Flow and Transport in Porous Media. Advances in Water Resources 34(9), 1102–1112 (2011)
12. Gersbacher, C.: The DUNE PrismGrid Module. In: Dedner, A., Flemisch, B., Klöfkorn, R. (eds.) Advances in DUNE, vol. 107, pp. 33–44. Springer, Heidelberg (2012)
13. Gräser, C., Sander, O.: The dune-subgrid module and some applications. Computing 8(4), 269–290 (2009)
14. Logg, A.: Automating the finite element method. Arch. Comput. Methods Eng. 14(2), 93–138 (2007)
15. MpCCI website, `http://www.mpcci.de`
16. Müthing, S.: dune-multidomain, `http://gitorious.org/dune-multidomain`
17. Müthing, S.: dune-multidomaingrid, `http://gitorious.org/dune-multidomaingrid`

Part II
External DUNE Modules

A Software Framework for Reduced Basis Methods Using DUNE-RB and RBMATLAB

Martin Drohmann, Bernard Haasdonk, Sven Kaulmann, and Mario Ohlberger

Abstract. Many applications from science and engineering are based on parametrized evolution equations and depend on time–consuming parameter studies or need to ensure critical constraints on the simulation time. For both settings, model order reduction by the reduced basis approach is a suitable means to reduce computational time. The method is based on a projection of an underlying high–dimensional numerical scheme onto a low–dimensional function space. In this contribution, a new software framework is introduced that allows fast development of reduced schemes for a large class of discretizations of evolution equations implemented in DUNE. The approach provides a strict separation of low–dimensional and high–dimensional computations, each implemented by its own software package RBMATLAB, respectively DUNE-RB. The functionality of the framework is exemplified for a finite–volume approximation of an instationary linear convection–diffusion problem.

1 Introduction

The reduced basis methods have gained increasing attention in recent years for stationary–elliptic, instationary–parabolic problems and various systems. In this contribution, we address the task of model reduction for parametrized scalar evolution equations. The functionality of the software framework described in this paper is restricted to linear problems, but during the development of the software

Martin Drohmann · Sven Kaulmann · Mario Ohlberger
Institute of Computational and Applied Mathematics, University of Münster, Einsteinstr. 62, 48149 Münster
e-mail: {mdrohmann,s_kaul01,ohlberger}@uni-muenster.de

Bernard Haasdonk
Institute of Applied Analysis and Numerical Simulation, University of Stuttgart, 70569 Stuttgart
e-mail: haasdonk@mathematik.uni-stuttgart.de

concept, we always bear in mind, that an extension to non–linear problems based on empirical operator interpolation is possible (cf. [3]).

Software engineering is a major problem for the development of reduced basis methods, because a reduced basis framework for a certain problem involves many steps, which can be expensive in both development and execution time. In this presentation, we show how to develop an extendible framework for linear evolution equations, based on our DUNE module DUNE-RB and the Matlab based software package RBMATLAB. This separation into two software packages serves two purposes: First, the use of a high-level programming language like Matlab with a vast library of mathematical methods accelerates the development and improvement of the abstract reduced basis methods, while the underlying discretizations can still be implemented efficiently in DUNE-RB. Second, it makes use of the decomposition of the reduced basis method into two computationally different phases, frequently denoted by the offline-/online-decomposition. During the computationally and time demanding *offline-phase* a low–dimensional function space, spanned by high–dimensional solution snapshots, is generated, which can be used to reduce the model order of the problem to only few degrees of freedom. Therefore, during the *online-phase*, reduced simulations can be computed in real-time and independently from the original discretization and high–dimensional model data. In this phase, an interactive user interface, like the one offered by Matlab, turns out to be a useful gift. We would like to mention, that a similar approach for finite–element discretizations based on the numerical software package *libmesh* has recently been developed for a variety of problems with an affine parameter dependence of the underlying data functions. [7]

Section 2 contains a very basic description of the reduced basis method for linear evolution equations. In sections 3.1 and 3.2 the two software packages RBMATLAB and DUNE-RB are introduced in detail, in terms of functionality and the specification of their interfaces. Finally, Section 4 elaborates on an example implementation for a finite–volume discretization of a linear convection–diffusion problem. We describe how to implement this scheme in DUNE-RB and how to reduce it in RBMATLAB.

2 Reduced Basis Method

In this section, we give a very short introduction to the reduced basis method for linear evolution equations. For a detailed description of the framework, including discussions on the efficiency of reduced simulations and the generation of reduced basis spaces by the "POD-Greedy" - algorithm, we refer to [6, 9].

The present study assumes problems of the following kind: For every parameter vector $\boldsymbol{\mu} \in \mathscr{D} \subset \mathbb{R}^p$, we are looking for a solution trajectory $u(\boldsymbol{\mu}) : [0, T_{\max}] \to \mathscr{W} \subset L^2(\Omega)$ fulfilling some linear evolution equation of the form

$$\partial_t u(t;\boldsymbol{\mu}) + \mathscr{L}[u(t;\boldsymbol{\mu})] = 0, \qquad u(0;\boldsymbol{\mu}) = u_0(\boldsymbol{\mu})$$

with suitable boundary conditions. Here $\mathscr{W} \subset L^2(\Omega)$ is a Hilbert space of functions on the spatial domain Ω and $\mathscr{L} : \mathscr{W} \to \mathbb{R}$ denotes a spacial differential operator.

There is a large number of available numerical schemes for solving problems of this kind. Many of those, which are frequently used, like finite–element, finite–volume or discontinuous Galerkin discretizations can be cast into a framework, where the solutions are approximated in a discrete function space $\mathscr{W}_h \subset \mathscr{W}$ of dimension H. First–order time discretizations then lead, for a given parameter vector $\boldsymbol{\mu}$, to a sequence of solution snapshots $u_h^k(\boldsymbol{\mu}) \in \mathscr{W}_h$ for $k = 1, \ldots, K$ computed by an initial projection of the initial data function

$$u_h^0(\boldsymbol{\mu}) = \mathscr{P}_h[u_0(\boldsymbol{\mu})] \tag{1a}$$

with a projection operator $\mathscr{P}_h : \mathscr{W} \to \mathscr{W}_h$ and iterative solutions of the equation

$$u_h^{k+1} + \Delta t^k \mathscr{L}_h^I(\boldsymbol{\mu}) \left[u_h^{k+1}(\boldsymbol{\mu})\right] = u_h^k(\boldsymbol{\mu}) - \Delta t^k \mathscr{L}_h^E(\boldsymbol{\mu}) \left[u_h^k(\boldsymbol{\mu})\right] \tag{1b}$$

for $k = 1, \ldots, K$ with suitable time–step sizes Δt^k. In the present study, we make the assumption that both the discrete operators $\mathscr{L}_h^I(\boldsymbol{\mu})$, $\mathscr{L}_h^E(\boldsymbol{\mu})$, the initial data function $\mathscr{P}_h[u_0(\boldsymbol{\mu})]$ and the constant function $b(\boldsymbol{\mu})$ depend affinely on the parameter, i.e. they can be written in separable forms

$$\begin{aligned} \mathscr{L}_h^I(\boldsymbol{\mu}) &= \sum_{q=0}^{Q_I} \sigma_I(\boldsymbol{\mu}) \mathscr{L}_h^{I,q}, & \mathscr{L}_h^E(\boldsymbol{\mu}) &= b(\boldsymbol{\mu}) + \sum_{q=0}^{Q_E} \sigma_E(\boldsymbol{\mu}) \mathscr{L}_h^{E,q}, \\ \mathscr{P}_h[u_0(\boldsymbol{\mu})] &= \sum_{q=0}^{Q_{u_0}} \sigma_{u_0}^q(\boldsymbol{\mu}) u_0^q, & b(\boldsymbol{\mu}) &= \sum_{q=0}^{Q_b} \sigma_b^q(\boldsymbol{\mu}) b^q, \end{aligned} \tag{2}$$

with parameter dependent coefficient functions $\sigma_I^q, \sigma_E^q, \sigma_{u_0}^q, \sigma_b^q : \mathscr{D} \to \mathbb{R}$, constant functions $u_0^q, b^q \in \mathscr{W}$ and linear operators $\mathscr{L}_h^{I,q}, \mathscr{L}_h^{E,q} : \mathscr{W} \to \mathscr{W}$ depending on the space variable only.

If we assume that there is an $N \ll H$ dimensional orthonormal basis $\Phi_N := \{\varphi_n\}_{n=1}^N$ for another discrete function space $\mathscr{W}_{\text{red}} \subset \mathscr{W}_h$ which somehow is a good approximation of the manifold of sought solutions $\{u_h^k(\boldsymbol{\mu}) \mid \boldsymbol{\mu} \in \mathscr{D}, k = 0, \ldots, K\}$, the numerical scheme (1a)-(1b) can be reduced by a Galerkin projection onto this so–called reduced basis space. For this, let $\mathscr{P}_{\text{red}} : \mathscr{W}_h \to \mathscr{W}_{\text{red}}$ be a Galerkin projection operator fulfilling the equation

$$\langle u - \mathscr{P}_{\text{red}}[u], \varphi \rangle = 0 \qquad \forall u \in \mathscr{W}_h \, \forall \varphi \in \mathscr{W}_{\text{red}}.$$

Then for each $\boldsymbol{\mu} \in \mathscr{D}$, sequences of low–dimensional solution snapshots $u_{\text{red}}^k(\boldsymbol{\mu}) \in \mathscr{W}_{\text{red}}$ for $k = 0, \ldots, K$ are obtained through the reduced numerical scheme

$$u_{\text{red}}^0(\boldsymbol{\mu}) = \mathscr{P}_{\text{red}} \circ \mathscr{P}_h[u_0(\boldsymbol{\mu})], \tag{3a}$$

$$u_{\text{red}}^{k+1} + \Delta t^k \mathscr{P}_{\text{red}} \circ \mathscr{L}_h^I \left[u_{\text{red}}^{k+1}(\boldsymbol{\mu})\right] = u_{\text{red}}^k(\boldsymbol{\mu}) - \Delta t^k \mathscr{P}_{\text{red}} \circ \mathscr{L}_h^E \left[u_{\text{red}}^k(\boldsymbol{\mu})\right]. \tag{3b}$$

These reduced solutions approximate the actual numerical solution in the higher dimensional discrete function space. In order to understand, why the above numerical scheme (3a)-(3b) can be computed efficiently, we switch to a vector based formulation of the numerical scheme by identifying the reduced solution sequences $u^k_{\text{red}}(\boldsymbol{\mu}) = \sum_{n=1}^N a^k_n(\boldsymbol{\mu})\varphi_n$ with their coefficient vectors

$$\mathbf{a}^k(\boldsymbol{\mu}) := \left(a^k_1(\boldsymbol{\mu}), \dots, a^k_N(\boldsymbol{\mu})\right)^t \in \mathbb{R}^N$$

and compute these vectors for each parameter vector $\boldsymbol{\mu} \in \mathscr{D}$ with the scheme

$$\mathbf{a}^0(\boldsymbol{\mu}) = \mathbf{u}_0(\boldsymbol{\mu}), \tag{4a}$$

$$\mathbf{a}^{k+1}(\boldsymbol{\mu}) + \Delta t \mathbf{L}^I(\boldsymbol{\mu})\left[\mathbf{a}^{k+1}(\boldsymbol{\mu})\right] = \mathbf{a}^k(\boldsymbol{\mu}) - \Delta t^k \mathbf{L}^E(\boldsymbol{\mu})\left[\mathbf{a}^k(\boldsymbol{\mu})\right], \tag{4b}$$

where the N-dimensional vectors $\mathbf{u}_0(\boldsymbol{\mu})$ and $\mathbf{b}(\boldsymbol{\mu})$ and the $N \times N$ sized matrices $\mathbf{L}^I$ and $\mathbf{L}^E$ are given by linear combinations of precomputed vectors or matrices, respectively.

$$\mathbf{L}^I(\boldsymbol{\mu}) = \sum_{q=1}^{Q_I} \sigma^q_I(\boldsymbol{\mu})\mathbf{L}^q_I, \qquad \mathbf{L}^E(\boldsymbol{\mu}) = \mathbf{b}(\boldsymbol{\mu}) + \sum_{q=1}^{Q_E} \sigma^q_E(\boldsymbol{\mu})\mathbf{L}^q_E,$$

$$\mathbf{u}_0(\boldsymbol{\mu}) = \mathbf{b}(\boldsymbol{\mu}) + \sum_{q=1}^{Q_{u_0}} \sigma^q_{u_0}(\boldsymbol{\mu})\mathbf{u}^q_0, \qquad \mathbf{b}(\boldsymbol{\mu}) = \sum_{q=1}^{Q_b} \sigma^q_b(\boldsymbol{\mu})\mathbf{b}^q.$$

The entries of the precomputed vectors $\mathbf{u}^q_0$ and $\mathbf{b}^q$ are computed as the projections onto the basis functions

$$\left(\mathbf{u}^q_0\right)_n := \int_\Omega u^q_0 \varphi_n, \qquad \left(\mathbf{b}^q_0\right)_n := \int_\Omega b^q \varphi_n \tag{5a}$$

and the matrices $\mathbf{L}^q_I$ and $\mathbf{L}^q_E$ are weighted gramian matrices with entries

$$\left(\mathbf{L}^q_I\right)_{nn'} := \int_\Omega \mathscr{L}^{I,q}_h[\varphi_{n'}]\,\varphi_n, \qquad \left(\mathbf{L}^q_E\right)_{nn'} := \int_\Omega \mathscr{L}^{E,q}_h[\varphi_{n'}]\,\varphi_n \tag{5b}$$

In order to gain control of the approximation error induced by the Galerkin projection onto reduced basis space, efficiently computable a posteriori error estimators are necessary. For linear problems, such estimators $\eta(\boldsymbol{\mu})$ assessing the error between reduced and detailed simulations $\max_{k=0}^K \left\| u^k_h(\boldsymbol{\mu}) - u^k_{\text{red}}(\boldsymbol{\mu})\right\| \le \eta(\boldsymbol{\mu})$ can be deduced in a way such that they integrate into the concept of offline-/online decomposition.

For details on a posteriori error estimators and the generation of a reduced basis space $\mathscr{W}_{\text{red}}$ with the so–called "POD–Greedy" algorithm, we refer to [6].

3 Software Concept

As mentioned in the introduction, the generation of the reduced basis method is split into two parts. The low–dimensional computations and abstract algorithms for the generation of the reduced basis methods are implemented in RBMATLAB, whereas DUNE-RB provides algorithms and data structures for the manipulation of the high–dimensional reduced basis space, and implements interfaces to several classes of parametrized high–dimensional problems, e.g. to linear evolution equations of the form (1a)-(1b). The two software packages can communicate by

1. either compiling DUNE-RB as a mex-library, which allows it to be called directly from the Matlab prompt, or
2. via TCP/IP communication over sockets.

Both methods rely on an encapsulation of Matlab data structures. For the first choice the memory handling of these data structures is kept in Matlab, whereas in the second case, data needs to be copied between both processes and re-instantiated on the DUNE-RB side as C++ data structures. Nevertheless, the second approach is assessed as more advantageous by the authors, because the compilation as a mex-library sets certain requirements on compilers and compiler options, as well as library dependencies. Furthermore, the TCP/IP interface allows to easily run the high–dimensional and the low–dimensional computations on different hardware platforms, each suitably chosen for its needs (server/client model).

In this section, the structure and the capabilities of both software packages are presented with a focus on the user interface methods for an already implemented numerical scheme. The implementation of a particular numerical scheme and the realization of a resulting reduced basis method for a parametrized linear evolution equation is then described in Section 4.

3.1 RBMATLAB

The Matlab based software package can be seen as a user interface for both the offline and the online phase of a reduced–basis–method execution cycle. Figure 1 illustrates the course of action for all reduced basis method implementations independent from the underlying problem. The headers in the boxes are code snippets as they are executed in RBMATLAB. Gray shaded boxes indicate interface functions where the arguments and return values are communicated to DUNE-RB via TCP/IP sockets and the actual computations are executed with DUNE. Steps 1 and 3 gather high–dimensional information needed for the generation of the reduced basis and step 4 finally reduces this data for use with efficient simulations by the reduced scheme. The actual implementation for these high–level functions is specified by the central object called `model`. The crucial attributes, this objects specifies, are *(i)* the problem type of the underlying parametrized partial differential equations and their discretizations, *(ii)* the chosen reduced basis generation strategy and *(iii)* the link to the numerical scheme in DUNE-RB.

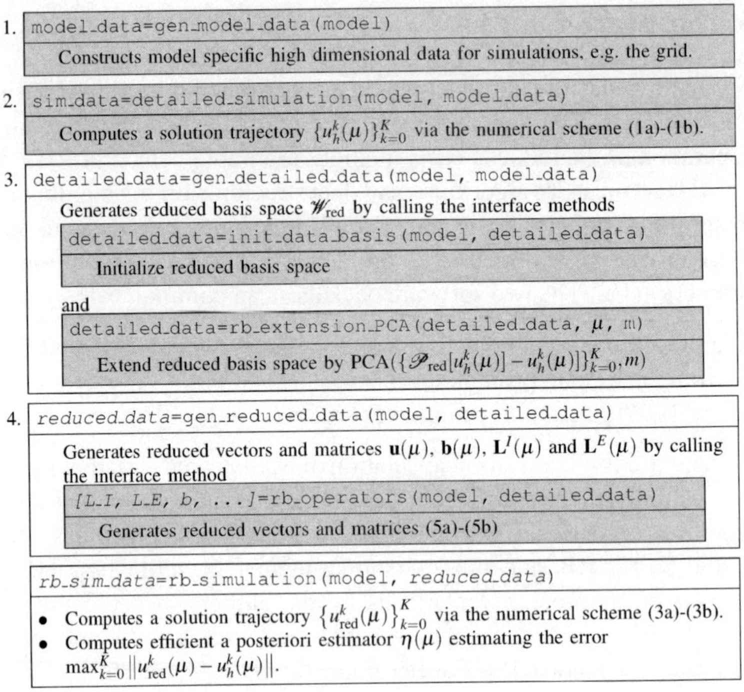

Fig. 1 Illustration of the course of action for reduced basis methods in RBMATLAB. Method calls delegated to DUNE-RB are surrounded with gray shaded boxes. The data structures `model_data`, `sim_data` and `detailed_data` are high–dimensional and depend on $\mathscr{O}(H)$, `reduced_data`, `[L_I, L_E, b, ...]` and `rb_sim_data` are low–dimensional data structures depending on $\mathscr{O}(N)$.

Apart from linear evolution problems of the form (1a)-(1b) there is also support for non–linear problems and discretization schemes where the data fields do not depend affinely on the parameter as required in (2). For this purpose, RBMATLAB provides the method of empirical operator interpolation proposed in [3].

For the reduced basis generation, RBMATLAB implements some variants, which are all an adaptation of the aforementioned "POD–Greedy"-algorithm. Noteworthy are the possibilities to adaptively refine the subset of the parameter space $\mathscr{D}$ from which the reduced basis functions are chosen [5] and to generate smaller reduced basis spaces for different subspaces of the parameter space [4].

For rapid prototyping, RBMATLAB contains a library of finite–volume discretizations for partial differential equations, which is based on triangular or rectangular grids in two dimensions and implements the interface for communication with the reduced basis algorithms in RBMATLAB. This interface can actually be implemented by arbitrary software packages providing numerical schemes for partial differential equations. In the next section, we focus on the software package DUNE-RB bringing the flexibility and power of DUNE to the reduced basis world.

3.2 DUNE-RB

The DUNE module DUNE-RB for the high–dimensional computations is based on the discretization module DUNE-FEM providing a layer of abstraction for discrete function spaces and differential operators [2]. DUNE-RB extends the concepts from DUNE-FEM and provides interface classes for *(i)* solution trajectories $\left\{u_h^k(\boldsymbol{\mu})\right\}_{k=0}^{K}$, *(ii)* reduced basis spaces $\mathscr{W}_{\mathrm{red}}$ and *(iii)* decomposed discrete operators of the form $\mathscr{L}_h[\cdot] = \sum_{q=1}^{Q} \sigma^q(\boldsymbol{\mu})\mathscr{L}_h^q[\cdot]$, where the notations are adopted from Section 2. This section gives an overview of the available implementations for these interfaces in DUNE-RB. In the end of this section, we elaborate on the implementation of the communication interface to RBMATLAB and further low–level functionality.

Interface classes: *(i)* The data structure for storing either a solution trajectory or reduced basis functions can in both cases be seen as a list of discrete functions from the high–dimensional function space $\mathscr{W}_h$. Such lists are given as implementations of an abstract `DiscreteFunctionListInterface` class. At the moment DUNE-RB provides the two implementations `DiscreteFunctionList_xdr` for efficient storage of the discrete functions on the harddrive in the xdr format and `DiscreteFunctionList_mem` holding all the discrete functions in main memory.

(ii) The reduced basis space class `ReducedBasisSpace` is derived from the DUNE-FEM interface `DiscreteFunctionSpaceInterface` with a base function list implemented as a `DiscreteFunctionListInterface`. In contrast to classical finite–element spaces implemented in DUNE-FEM, this class offers additional methods for manipulating the space by adding and changing global basis functions.

(iii) DUNE-FEM introduces the concept of local operators which are defined by iteratively applying the operator to local representations of the argument's or destination's functions living on a grid entity and its neighbors only. This concept allows to implement matrix–free discrete operators saving memory and bandwidth, and to concatenate these operators without the overhead of further grid iterations, which are very expensive at least for unstructured grids. DUNE-RB extends this concept by `LocalParametrizedOperatorInterface` classes, enriching the local operator implementation by the method `coefficient()` scaling the operator's output by the parameter dependent coefficient function. A sum of such parametrized and locally implemented operators is managed by decomposed operators implementing the `DecomposedOperatorInterface`. Such instances are actually discrete operators by their own, evaluating to the entire sum, but also providing access to the coefficients– and parameter–independent operator parts, which are needed for construction of reduced matrices.

Decomposed operators are implemented for general finite–volume operators for various numerical flux implementations, like the Lax–Friedrichs flux. The decomposition of these operators is derived automatically from the affinely parameter dependent decomposition of the used data functions. This will be described in more detail, in Section 4.2.

Communication interface: The communication interface to RBMATLAB is simply achieved by instantiating an `RBSocksServer` object provided with a template parameter derived from `RBMatlabBase`. The class given as the template parameter needs to call the derived method `registerFunction("op", mp)` for each operation `"op"` that shall be available for triggering from the other side, binding it to a program entry point given by the second argument `mp`. Calling the method `run()` on the server object then puts the program into "listening" mode awaiting orders from RBMATLAB.

Furthermore, this DUNE module provides a singleton class `Parameter`, which globally manages the parameter vector $\boldsymbol{\mu} \in \mathscr{D}$ in such a way that variables depending on the parameter vector simply delegate the evaluation of these magnitudes to the `Parameter` singleton instance.

4 Example: Linear Evolution Equation

In this section, we analyze the reduced basis method for a linear instationary convection–diffusion problem implemented in DUNE for two and three dimensions. The considered problem looks as follows: For each parameter vector $\boldsymbol{\mu} \in \mathscr{D}$, we are looking for solutions $u(t;\boldsymbol{\mu}) \in BV(\Omega) \cap L^\infty(\Omega) \subset L^2(\Omega)$ fulfilling the equations

$$\partial_t u(t;\boldsymbol{\mu}) + \mathbf{v}(\boldsymbol{\mu})\nabla u(t;\boldsymbol{\mu}) - D(\boldsymbol{\mu})\Delta u(t;\boldsymbol{\mu}) = 0 \qquad \text{in } \Omega \times [0, T_{\max}] \tag{6a}$$

$$u(0;\boldsymbol{\mu}) = u_0 \qquad \text{in } \Omega \times \{0\} \tag{6b}$$

$$u(t;\boldsymbol{\mu}) = u_{\text{dir}} \qquad \text{on } \Gamma_{\text{dir}} \times [0, T_{\max}] \tag{6c}$$

$$(\mathbf{v}(\boldsymbol{\mu})\nabla u(t;\boldsymbol{\mu})) \cdot \mathbf{n} = 1 \qquad \text{on } \Gamma_{\text{neu}} \times [0, 1] \tag{6d}$$

on a rectangular domain $\Omega := [0,1]^d$ with $d \in \{2,3\}$ and $T_{\max} = 1.0$. In our computations, we consider a three dimensional parameter space $\mathscr{D} := [0, 0.001] \times [0.3, 1] \times [0.3, 1]$ and the parametrized data functions are given by $D(x;\boldsymbol{\mu}) = \mu_1$ and $\mathbf{v}(\boldsymbol{\mu}) = (\mu_2, \mu_3, 0.1)$. The Dirichlet boundary function is given by $u_{\text{dir}} = 0.01$ and the initial data function $u_0(x) = \exp(-10\|x - 0.5\|_2)\chi_{\|x-0.5\|_2 \leq 0.2}$ implements a non–smooth circle–shaped concentration. The domain's boundary is separated into an "inflow" boundary $\Gamma_{\text{dir}} := \{0\} \times [0,1] \cup [0,1] \times \{0\}$ with a Dirichlet condition and an "outflow" Neumann boundary $\Gamma_{\text{neu}} := \partial\Omega \backslash \Gamma_{\text{dir}}$.

4.1 *Finite-Volume Discretization*

The problem is discretized with a purely explicit finite–volume scheme, where the convection part is discretized by a Lax–Friedrichs flux. For a discretization of the form (1a)-(1b), we therefore need to specify the operator $\mathscr{L}_h^E(\boldsymbol{\mu})$. In our example, the implicit operator $\mathscr{L}_h^I(\boldsymbol{\mu})$ is set to zero. For the time discretization we choose a global time–step size Δt, which is small enough such that a CFL–type condition is fulfilled for all parameters $\boldsymbol{\mu} \in \mathscr{D}$. Before we define the implemented discretization operator, however, we need to fix some notations concerning the tessellation of the domain: $\mathscr{T} := \{e_i\}_{i=1}^H$ denotes a numerical grid consisting of H disjoint polygonal elements forming a partition of the domain $\bar{\Omega} = \bigcup_{i=1}^H \bar{e}_i$. For each element $e_i, i = 1,\dots,H$, we assume that there exist

- certain points x_i lying inside the element e_i, such that points in adjacent elements are perpendicular to the corresponding edges and
- a set of indices $\mathscr{N}(i) := \mathscr{N}_{\text{in}}(i) \cup \mathscr{N}_{\text{dir}}(i) \cup \mathscr{N}_{\text{neu}}(i)$ counting the element's edges e_{ij} for $j \in \mathscr{N}(i)$, where $\mathscr{N}_{\text{in}}$ corresponds to edges between inner elements of the domain or elements adjacent by cyclical boundary conditions, $\mathscr{N}_{\text{dir}}$ includes those edges on the Dirichlet boundary and $\mathscr{N}_{\text{neu}}$ those ones on the Neumann boundary of the domain.

On each edge e_{ij}, we denote their barycenters by x_{ij} and their outer unit normals by $\mathbf{n}_{ij}$.

Furthermore, we define the finite–volume space $\mathscr{W}_h$ as the span of base functions $\psi_i \in \mathscr{W}, i = 1,\dots,H$ being piecewise constant on the i-th cell e_i and vanishing elsewhere. For a function $u_h \in \mathscr{W}_h$ its degrees of freedom (DoFs) are given by $u_{h,i} = u_h(x_i)$, allowing a DoF–wise definition of the explicit discretization operator

$$
\begin{aligned}
\left(\mathscr{L}_h^E(\boldsymbol{\mu})\,[u_h]\right)_i = & \frac{1}{|e_i|} \sum_{j \in \mathscr{N}_{\text{in}}(i)} g_{ij}^{\text{lf}}(u_{h,i}, u_{h,j}) + g_{ij}^{\text{diff}}(u_{h,i}, u_{h,j}) \\
& + \frac{1}{|e_i|} \sum_{j \in \mathscr{N}_{\text{dir}}(i)} g_{ij}^{\text{lf}}(u_{h,i}, u_{\text{dir}}(x_{ij})) + g_{ij}^{\text{diff}}(u_{h,i}, u_{\text{dir}}(x_{ij})) \\
& + \frac{1}{|e_i|} \sum_{j \in \mathscr{N}_{\text{neu}}(i)} \int_{e_i} u_{\text{neu}}
\end{aligned} \tag{7}
$$

with numerical fluxes, c.f. [8]

$$
g_{ij}^{\text{lf}}(u,v) := \frac{1}{2}\left(\mathbf{v}(\boldsymbol{\mu}) \cdot \mathbf{n}_{ij}(u+v) + \frac{1}{\lambda}(u-v)\right) \text{ and} \tag{8}
$$

$$
g_{ij}^{\text{diff}}(u,v) := \frac{|e_{ij}|}{|x_i - x_j|} D(x_{ij};\boldsymbol{\mu})(u-v). \tag{9}
$$

Remark 1. In order to fulfill the requirements on the separable form of the operator as stated in (2), it suffices to require the data functions u_0, u_{dir}, D and $\mathbf{v}$ to be affinely parameter dependent. It can be easily observed that a decomposition into a sum of products of purely parameter and space dependent data functions is inherited by these linear numerical fluxes and, therefore, leads to the desired decomposition of the discrete operator.

4.2 Implementation

A major design principle of DUNE-RB is to enable quick development of new numerical schemes without knowledge of the reduced basis method framework. For this purpose, DUNE-RB provides code snippets for easy construction of programs run as a TCP/IP server controlled by RBMATLAB.

Listing 1 (Excerpt from a typical main file)

```
typedef DefaultDescr
            < LinEvolExplicitDiscretization,
              ConvDiffModel >                   Description;

typedef ProblemTraits
           < LinEvolDefault, LinEvolFacade,
             RBSocksServer, Description > ServerTraits;


%#include <dune/rb/matlab/duneserver/main.inc>
#include <main.inc>
```

Listing 1 shows an example code file for such a server, where the developer only needs to provide two Traits class declarations. The `ServerTraits` class provides the full declaration of the server by specifying the interface methods made available in RBMATLAB (in this case `LinEvolFacade`) and the actual entry points of these commands (`LinEvolDefault`). The latter one, of course, depends on the problem definition and the discretization method, specified by the typedef `Description` This class again consists of

- an abstract problem type `LinEvolExplicitDiscretization` implementing the numerical scheme (1a)-(1b) in this case, and
- the declaration for the spatial operators $\mathscr{L}_h^I(\boldsymbol{\mu})$ and $\mathscr{L}_h^E(\boldsymbol{\mu})$ provided by the model class `ConvDiffModel`.

As described in Remark 1, the concept of inheriting the decomposed structure of discrete operators from data functions is also implemented in DUNE-RB, which further simplifies the definition of the aforementioned model class.

4.3 Results

In Figure 2 we demonstrate the behavior of the error estimator during the construction of the reduced basis via the "POD–Greedy" algorithm. Plotting the error bound in logarithmic scale against the size of the reduced basis, varying from one to 30, we are able to see an approximately exponential decay of the error, as expected from the theory.

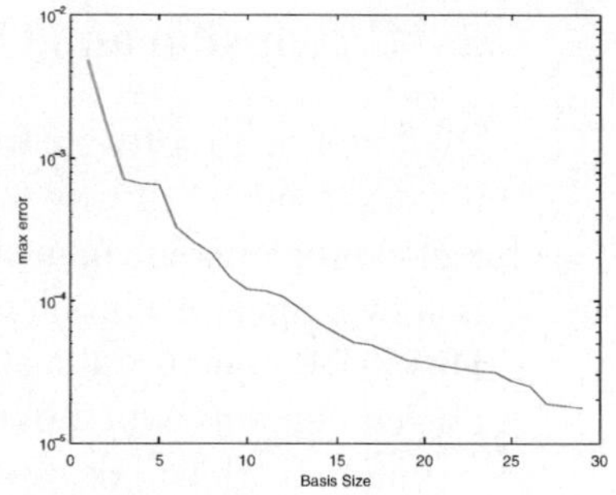

Fig. 2 Maximum of error bound over a set of 216 parameters for different reduced basis sizes in logarithmic scale.

Table 1 Numerical results for a transport problem in 2D and 3D with non–divergence–free velocity. The given error is the mean error between full and reduced simulation, tested with 250 randomized parameter values

	High-dim. Solution (s)	RB Generation (s)	Gen. of Online Matrices (s)	Reduced Sim. (s)	Reconstruction (s)	Grid Cells	$L^\infty - L^2$-Error
2D Transport (25 Base Functions)	11	71	6.69	0.11	0.33	1,024	$1.42 \cdot 10^{-3}$
2D Transport (50 Base Functions)	11	2,250	21	0.15	0.42	1,024	$4.64 \cdot 10^{-4}$
3D Transport (50 Base Functions)	944	$1.57 \cdot 10^5$	4,659	0.15	26	32,768	$9.11 \cdot 10^{-4}$

In Table 1 we present runtimes for the model problem described in Section 4 in two and three space dimensions as generated by our implementation. Column one gives the time for the computation of one trajectory using a `YaspGrid` as a grid manager with the numbers of cells as indicated in column six. Columns two and three give the important numbers for the offline part of the reduced basis algorithm: The time for the "POD–Greedy" algorithm and the Galerkin projection step, that is, the computation of the low–dimensional matrices. In columns four and five we see the runtimes for the online part of the RB method: the solution of the reduced system and the reconstruction of a high–dimensional function from a reduced one, respectively.

With increasing size of the reduced basis from row one to two, and as well, with growing world dimensions, one can observe a significant growth in the runtime for the offline algorithm. Still, the rather time consuming offline phase, with a total runtime of about 45 hours in the 3D case, pays off with a speedup–factor of about 25 and 20 in the 2D cases and even 36 for the 3D case (online phase of the RB method compared to one high–dimensional solution). All these factors consider the reconstruction step to be part of the online phase, which is not the case if one is only interested in the value of an output functional, for example. Considering only the run–times for one high– and ones low–dimensional solution, the speedup–factor grows to 6293 in the 3D case. In all these cases we deal with a relative error of 10^{-3} to 10^{-4}, which is quite acceptable.

5 Conclusion and Outlook

We developed a software framework for the reduced basis methods simplifying both the development of new reduced basis algorithms and the implementation of new model problems and numerical schemes. An example for a finite–volume discretization of a linear evolution equation has been implemented in our software module DUNE-RB demonstrating the flexibility of DUNE by using the same model description on domains with different dimensions.

Future work will deal with the implementation of a library of more complex test cases and the extension of the software framework to non–linear problems and to systems of parametrized partial differential equations. Furthermore, new ideas for the generation of reduced basis spaces will be implemented in RBMATLAB and evaluated with the aforementioned test cases. Theoretically, most of the algorithms implemented in DUNE-RB can be run in parallel because of the parallel capabilities of DUNE, but practical tests of this functionality are still outstanding.

Finally, we plan to publish both software packages with installation instructions and documentation on our project homepage `http://morepas.org`

References

1. Barrault, M., Maday, Y., Nguyen, N., Patera, A.: An 'empirical interpolation' method: application to efficient reduced-basis discretization of partial differential equations. C. R. Math. Acad. Sci. Paris Series I 339, 667–672 (2004)
2. Dedner, A., Klöfkorn, R., Nolte, M., Ohlberger, M.: A generic interface for parallel and adaptive discretization schemes: abstraction principles and the DUNE-FEM module. Computing 90, 165–196 (2010)
3. Drohmann, M., Haasdonk, B., Ohlberger, M.: Reduced Basis Approximation for Nonlinear Parametrized Evolution Equations based on Empirical Operator Interpolation. SIAM J. Sci. Comput. 34(2), 937–969 (2012)
4. Haasdonk, B., Dihlmann, M., Ohlberger, M.: A training set and multiple bases generation approach for parametrized model reduction based on adaptive grids in parameter space. Math. Comput. Model. Dyn. Syst. 17(4), 423–442 (2011)
5. Haasdonk, B., Ohlberger, M.: Adaptive basis enrichment for the reduced basis method applied to finite volume schemes. In: Proc. 5th International Symposium on Finite Volumes for Complex Applications, pp. 471–478 (2008)
6. Haasdonk, B., Ohlberger, M.: Reduced basis method for finite volume approximations of parametrized linear evolution equations. M2AN, Math. Model. Numer. Anal. 42(2), 277–302 (2008)
7. Knezevic, D., Petterson, J.: A high-performance parallel implementation of the certified reduced basis method. Comput. Meth. Appl. Mech. Eng. 200, 1455–1466 (2011)
8. Kröner, D.: Numerical Schemes for Conservation Laws. John Wiley & Sons and Teubner (1997)
9. Patera, A., Rozza, G.: Reduced Basis Approximation and a Posteriori Error Estimation for Parametrized Partial Differential Equations. MIT (2007), `http://augustine.mit.edu/methodology/methodology_bookPartI.htm`; Version 1.0, Copyright MIT 2006-2007, to appear in (tentative rubric) MIT Pappalardo Graduate Monographs in Mechanical Engineering

DUNE-UDG: A Cut-Cell Framework for Unfitted Discontinuous Galerkin Methods

Christian Engwer* and Felix Heimann

Abstract. Simulations on complex shaped domains are of big interest as the mesh generation for such domains is still an involved process. Recently, cut-cell based methods are becoming very popular. These methods avoid the problems of mesh-generation by using unfitted discretizations on cut-cell meshes. We present the DUNE-UDG module which allows an easy implementation of Unfitted Discontinuous Galerkin methods on cut-cell grids. Different geometry representations are available. Using the presented interfaces it is possible to implement completely new cut-cell representations with a minimum of work.

1 Introduction

The computational efficiency of many numerical methods may be increased significantly if their spatial discretization corresponds to structured cartesian grids. Cutting the solution's domain out of a structured grid, emerges as an obvious approach to reconcile the needs of applications in complicated domains with the needs of efficient implementations. Due to the inherent geometric implications, Riemann solvers and Finite Volume discretizations for elliptic problems were the first to be employed in such *cut-cell* or *adaptive cartesian* schemes [7, 16]. Recently, similar ideas were also used in the context of finite element methods, e.g. [17, 12].

The authors of [3, 11, 13] showed the applicability of *Unfitted Discontinuous Galerkin* (UDG) methods on cut-cell grids for elliptic, parabolic and incompressible viscous flow problems. The steady and synchronous development of such a vast range of different methods in the context of numerical sciences requires a software interface which is sufficiently unified to be simultaneously employed by all

Christian Engwer · Felix Heimann
IWR, Univerität Heidelberg, INF 368, 69120 Heidelberg
e-mail: `{felix.heimann, christian.engwer}@iwr.uni-heidelberg.de`

* Now INAM, WWU Münster.

applications but still flexible enough to allow for experimental implementations and rapid prototyping.

We present the DUNE-UDG module, a software solution which meets these demanding requirements. It is based on the conceptual description of cut-cell methods as given in Section 2. The corresponding interface is discussed in Section 3. Although designed for UDG discretizations, the provided functionality allows the application of the original Finite Volume cut-cell methods (which require only a subset of the interface) as well as the integration with related approaches such as the *extended Finite Elements* method [8, 5]. Example applications of UDG methods will be shown in Section 4.

2 Concepts

2.1 The Unfitted Discontinuous Galerkin Method

In [3] the authors describe an Unfitted Discontinuous Galerkin (UDG) method for solving PDEs in complex domains. It is based on the idea of Unfitted Finite Elements, see [1], but uses a Discontinuous Galerkin formulation to weakly enforce boundary conditions along the unfitted domain boundary. This approach allows the use of higher order trial and test functions. For problems described by a conservation law, DG methods are especially attractive as many DG formulations are locally mass conservative and therefore able to accurately describe fluxes over element boundaries.

Given a domain $\Omega \subseteq \mathbb{R}^d$, $\mathscr{G}$ describes a disjoint partitioning into N subdomains

$$\mathscr{G}(\Omega) = \left\{\Omega^{(0)}, \ldots, \Omega^{(N-1)}\right\}. \tag{1}$$

The partitioning $\mathscr{G}$ is usually based on geometrical properties obtained from experiments or previous simulations. In general the boundaries $\partial\Omega^{(i)}$ exhibit a complex shape.

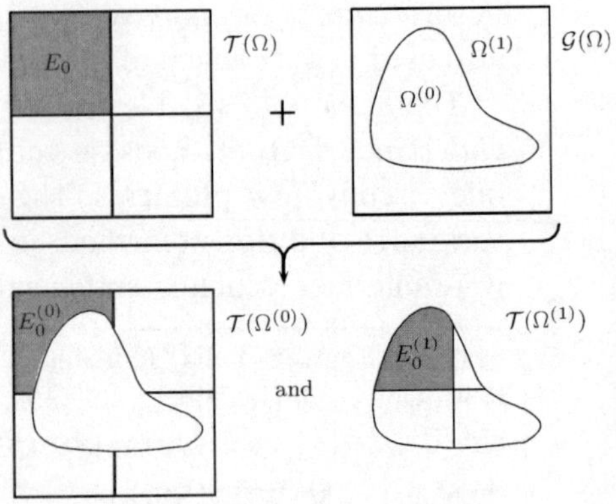

Fig. 1 Construction of the partitions $\mathscr{T}(\Omega^{(i)})$ given the partitions $\mathscr{G}$ and $\mathscr{T}$ of the domain Ω.

The Finite Element Mesh for a subdomain $\Omega^{(i)}$ is based on a fundamental mesh

$$\mathscr{T}(\Omega) = \{E_0, \ldots, E_{M-1}\} \tag{2}$$

of the whole domain Ω and is defined as the intersection of $\Omega^{(i)}$ and the fundamental mesh, see Fig. 1:

$$\mathscr{T}(\Omega^{(i)}) = \left\{ E_n^{(i)} = \Omega^{(i)} \cap E_n \,\middle|\, E_n^{(i)} \neq \emptyset \right\} . \tag{3}$$

Note that the elements $E_n^{(i)}$ can be arbitrarily shaped and in general will not be convex.

On the fundamental element $E_n \in \mathscr{T}(\Omega)$, $\varphi_{n,j} \in P_k$ denotes a polynomial, where P_k is the space of polynomial functions of degree k. The shape functions $\varphi_{n,j}^{(i)}$ are given by polynomials $\varphi_{n,j} \in P_k$ with their support restricted to $E_n^{(i)} \in \mathscr{T}(\Omega^{(i)})$. The resulting finite element space is defined by $V_k^{(i)} = \{v \in L_2(\Omega^{(i)}) | \; v|_{E_n^{(i)}} \in P_k\}$ and is discontinuous on the between elements. The discontinuity of a function $x \in V_k^{(i)}$ on the interface between two adjacent elements $E_n^{(i)}$ and $E_m^{(i)}$ is defined as $[\![x]\!] = x|_{\partial E_n^{(i)}} \mathbf{n}_{E_n^{(i)}} + x|_{\partial E_m^{(i)}} \mathbf{n}_{E_m^{(i)}}$. On the boundary $\partial\Omega^{(i)}$ we define the jump $[\![x]\!] = x\mathbf{n}_{E_n^{(i)}}$. The average of $x \in V_k^{(i)}$ is defined as $\{x\} = \frac{1}{2}(x|_{\partial E_n^{(i)}} + x|_{\partial E_m^{(i)}})$ on the interface and as $\{x\} = x$ on $\partial\Omega^{(i)}$.

The UDG method has been applied successfully to a range of different problems. Being first published for an elliptic model problem [3], the method has been proven of value also for time-dependent problems, including problems with time dependent domains, see [10, 11, 13]. Implementations for other DG schemes can easily be added in their primal formulation as the subdomains and unfitted boundaries are completely handled by the assembler and special quadrature rules.

2.2 The Cut-Cell Mesh

Assembling the local stiffness matrix in a DG approach requires integration over the volume of each element $E_n^{(i)}$ and its surface $\partial E_n^{(i)}$. Mesh elements obtained by the UDG method might exhibit very complicated shapes, thus quadrature rules based on interpolation functions are not directly applicable.

In order to guarantee accurate evaluation of integrals in an efficient manner, quadrature rules for irregular shaped elements are constructed using a local triangulation of $E_n^{(i)}$. To do so, $E_n^{(i)}$ is subdivided into a disjoint set $\{E_{n,k}^{(i)}\}$ of simple geometric objects, i.e. simplices and hypercubes. In the following, we call $E_{n,k}^{(i)}$ an *integration part*, each can be integrated using efficient Gauss type quadrature rules. The construction of the local triangulation depends on the representation of the geometry. Here, different algorithms are possible and available in the library, see [9].

As noted before, the cut-cell elements $E_n^{(i)}$ can be arbitrarily shaped. A particular problem is that the cell volume can be arbitrarily small, $|E_n^{(i)}| \ll |E_n|$.

The one problem is, that such small cells lead to a bad condition of the stiffness matrix and to nearly decoupled cells. Let's consider monomial shape-functions and a cut-cell covering the lower left corner. On the fundamental cell E_n the shape-functions cover the range of $[0,1]$, while on the cut cell $E_n^{(i)}$ the range much smaller $[0,\varepsilon]$. This problem is solved by an appropriate scaling of the shape-functions. Instead of scaling them relative to the fundamental cell, they are scaled according to the bounding box of the cut-cell. This is also helpful in the case of anisotropic elements.

Special means have to be applied in case of time dependent simulations using explicit time stepping methods. As the cut cells may be arbitrarily small, they lead to very strict time step stability restrictions. A common approach to overcome these constraints is a cell merging technique. The general idea is to merge small cut cells (fragment cells) into larger neighboring cells (mother cells). In purely finite volume simulations, it is sufficient to only merge the fluxes of the fragment cell with those of the mother cell [7]. For the UDG approach, it is necessary to construct the complete merged geometry, in order to correctly evaluate volume integrals. In the literature, different approaches are proposed on how to identify fragment cells and how to choose the mother cell for the merging procedure. In the presented library, we allow to use different merging strategies, depending on the problem.

As described in [2] DUNE does not allow to store data directly in the mesh. Instead the user can associate data with certain entities, using an `IndexSet`. An `IndexSet` offers a consecutive, zero starting numbering of all entities of the same `GeometryType` and `codimension`. Not all entities of the fundamental mesh $\mathcal{T}(\Omega)$ are contained in the triangulation $\mathcal{T}(\Omega^{(i)})$ of the i'th subdomain; some are lying outside the subdomain and some got merged with neighboring cells. Thus it is not possible to use the `IndexSets` of the fundamental mesh anymore. In order to associate data with the cells in a subdomain mesh $\mathcal{T}(\Omega^{(i)})$, DUNE-UDG offers a consecutive numbering of the cells in $\mathcal{T}(\Omega^{(i)})$, following the concept of `IndexSets` in DUNE-GRID. These indices exist for entities of codimension 0, they are consecutive and zero starting within each subdomain and allow to construct arbitrary `Mappers` depending on which equations are to be solved in which subdomain.

3 Implementation

The described concepts of the UDG method are implemented in the DUNE-UDG module. It is built on DUNE-GRID and DUNE-COMMON to provide the infrastructure to compute integration parts on a cut-cell mesh. Different implementations of geometry representation and appropriate local triangulation algorithms are available.

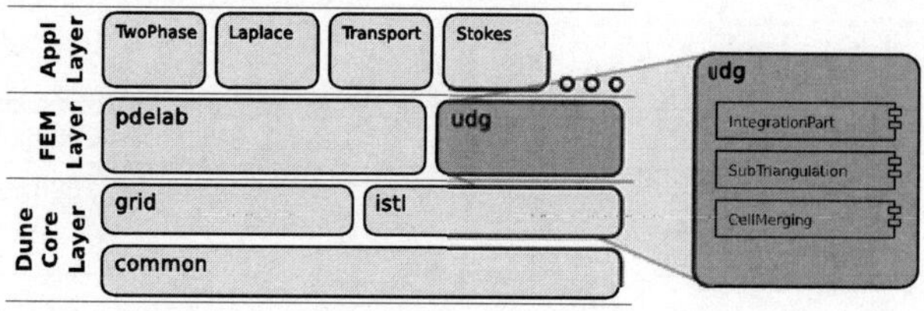

Fig. 2 Integration of the DUNE-UDG module with the DUNE framework.

DUNE-UDG itself offers a limited discretization interface. To perform complex multi-physics computations, it integrates with DUNE-PDELAB. It is possible to use any Discontinuous Galerkin local operator available in DUNE-PDELAB on a DUNE-UDG cut-cell mesh.

In the following we will introduce the most important class of DUNE-UDG and describe how they implement the concepts of Section 2. All classes belonging to the DUNE-UDG module are encapsulated in a namespace `Dune::UDG`: i) the `VirtualSubTriangulation` computes the local triangulation of fundamental cells, ii) Integration parts of cell and intersections are represented in the `EntityPart` and `IntersectionPart` classes, iii) using `CellMerging` fragment cells can be merged with neighboring cells and iv) the `IndexSet` allows to associate data with cells contained in given subdomain.

3.1 VirtualSubTriangulation

The `Dune::UDG::VirtualSubTriangulation` interface represents the functionality of the local triangulation for one or more subdomains. Three different implementations of the `Dune::UDG::VirtualSubTriangulation` interface are available:

`Dune::SubTriangulation2D` Implementation of a two-dimensional subdomain. The domain boundary is given by a list of geometric primitives.

`Dune::UDG::MarchingCubeSubTriangulation` A subdomain in two or three dimensions is given by a scalar function. This class implements the local triangulation of implicitly described geometries.

`Dune::UDG::NoSubTriangulation` This implementation assumes $\Omega^{(0)} = \Omega$ and does not perform any local triangulation. It is used for debugging.

All implementations are parameterized by a fundamental mesh and a geometry description. The fundamental mesh is given as a `Dune::GridView` object. Description of the geometry does vary between the different implementations. The evaluation of volume and surface integrals is implemented using the methods `create_entity_parts` and `create_intersection_parts`, which generate sets of sub-elements. Note that these methods return a list of integration parts for all domains intersecting with a given entity E_n on the fundamental mesh. Additionally `Dune::UDG::VirtualSubTriangulation` exposes a method

Table 1 The `Dune::UDG::VirtualSubTriangulation` interface.

Member	Description
`GridView`	type of the fundamental mesh GridView
`EntityPointer, Entity`	types of the codimension 0 EntityPointer and Entity on the fundamental mesh
`EntityPart`	type for a codimension 0 integration part $E_{n,k}^{(i)}$
`IntersectionPart`	type for a codimension 1 integration part
`BoundingBox`	type for the bounding box of $E_n^{(i)}$
`create_entity_parts(EntityPointer, std::list<EntityPart>)`	calculate the list of codimension 0 integration parts of E_n
`create_intersection_parts(EntityPointer, std::list<IntersectionPart>)`	calculate the list of codimension 1 integration parts of E_n
`BoundingBox boundingBox(Entity, domain_index)`	calculate bounding box for $E_n^{(i)}$; required for the scaling of local basis functions
`bool isInDomain(Entity, domain_index)`	test whether a given entity E_n in the fundamental mesh is contained in a certain subdomain
`ctype getEdgeNorm(IntersectionPart)`	obtain the sum over all integration parts of the intersection, a given part belongs to.

`boundingBox` to compute the bounding box of an element. During evaluation of the local basis functions the bounding box is necessary for the correct scaling of the basis functions.

The `Dune::UDG::MarchingCubeSubTriangulation` implementation allows a highly intuitive definition of multiple domains via I continuous scalar level set functions ϕ_i, $(i = 1 \dots I)$. Given a point $x \in \Omega$, each function gives either $\phi_i(x) \geq 0$ or $\phi_i(x) < 0$. This binary property may be utilized to define 2^I subdomains Ω_j, $(j = 1 \dots 2^I)$ with boundaries $\partial\Omega_j$ fulfilling

$$\forall x \in \partial\Omega_j : \qquad x \in \partial\Omega \quad \vee \quad \exists\, i \in \{1,\dots,I\} : \phi_i(x) = 0.$$

To construct the corresponding sub triangulation, an element or intersection of the fundamental mesh is dissected with regard to the function values of ϕ_1 according to the algorithm in [3]. Afterwards, the generated entity parts or intersection parts are recursively processed with the same algorithm for each ϕ_i with $1 < i \leq I$. Due to its simplicity and sequential nature, this approach does not result in any restrictions regarding the topological properties of the subdomains.

3.2 *Integration Parts*

The integration parts, as they are constructed by the `Dune::UDG::VirtualSubTriangulation`, are represented by the classes `EntityPart` and `IntersectionPart`. These are modeled similar to the `Dune::Intersection` of DUNE-GRID. They provide geometric mappings from the integration part to the associated fundamental cells, i.e. their bounding boxes, as well as mappings to world coordinates.

Table 2 Members of the `Dune::UDG::EntityPart`.

Member	Description
`GlobalGeometry`	type of the mapping to global coordinates
`LocalGeometry`	type of the mapping to the bounding box
`GlobalGeometry& geometry()`	return the mapping from local coordinates on the integration part to global coordinates
`LocalGeometry& geometryInCell()`	return the mapping from local coordinates on the integration part to the bounding box of the fundamental cell
`int domainIndex()`	return the index of the domain this part belongs to.

The class `Dune::UDG::EntityPart` represents an integration part of a codimension 0 entity $E_n^{(i)} \in \mathcal{T}(\Omega^{(i)})$. It provides mappings from local coordinates of the integration part to global coordinates and to local coordinates in the bounding box of the fundamental mesh cell. Furthermore, the method `domainIndex` gives the index i of the subdomain $\Omega^{(i)}$ the integration part belongs to. Together with this `domainIndex` and the entity E_n of the fundamental mesh the user can find the associated degrees of freedom.

Table 3 Members of the `Dune::UDG::IntersectionPart`.

Member	Description
`GlobalGeometry`	type of the mapping to global coordinates
`EntityPoint`	type of the codimension 0 EntityPointer of the fundamental mesh
`LocalGeometry`	type of the mapping to a bounding box
`GlobalCoordinate`	type of world coordinates
`LocalCoordinate`	type of local coordinates on the integration part
`bool boundary()`	returns true if this intersection is part of a subdomain boundary $\partial\Omega^{(i)}$
`bool internalBoundary()`	returns true if this intersection is part of an internal boundary $\partial\bar{\Omega}^{(i)} \cap \partial\bar{\Omega}^{(j)}$
`bool neighbor()`	returns true if a neighboring cell exists
`EntityPointer inside()`	get access to the inside fundamental mesh cell
`EntityPointer outside()`	get access to the outside fundamental mesh cell
`GlobalGeometry& geometry()`	return the mapping from local coordinates on the integration part to global coordinates
`LocalGeometry& geometryInInside()`	return the mapping from local coordinates on the integration part to the bounding box of the inside fundamental cell
`LocalGeometry& geometryInOutside()`	return the mapping from local coordinates on the integration part to the bounding box of the outside fundamental cell
`GlobalCoordinate unitOuterNormal(LocalCoordinate)`	return the unit outer normal vector
`int boundarySegmentIndex()`	the macro grid boundarySegmentIndex of the fundamental mesh.
`int insideDomainIndex()`	the index of the domain the inside cell belongs to.
`int outsideDomainIndex()`	the index of the domain the outside cell belongs to.

Codimension 1 integration parts are represented by the `Dune::UDG::IntersectionPart` class. This class is very similar to the `Dune::Intersection` of DUNE-GRID and implements most of its interface. In contrast to a grid intersection, it can also model the intersection between cells of different subdomains.

Thus it provides domain information about the neighboring cells via the methods `insideDomainIndex` and `outsideDomainIndex`.

As for `Dune::Intersection` the method `boundary` returns true if the intersection is part of a boundary. This can be either a boundary of the fundamental mesh $\partial\Omega$ or an internal boundary $\partial\bar{\Omega}^{(i)} \cap \partial\bar{\Omega}^{(j)}$ between two subdomains. In order to identify the latter, `Dune::UDG::IntersectionPart` has an additional method `internalBoundary`.

The `Dune::UDG::BoundingBox` class represents a cartesian bounding box of a subdomain cell $E_n^{(i)}$. It provides a reduced `Dune::Geometry` interface and gives access to the fundamental mesh cell E_n and the domain index i, using the methods `entity` and `domainIndex`.

4 Applications

In this Section, we present two example applications using DUNE-UDG. The first one shows the solution of the Stokes equation on a complicated domain. The second example shows instationary two-phase-flow computations, where the domain itself is changing in time.

4.1 *Pore-Scale Stokes*

We consider laminar flow through a porous medium. The viscous flow is modeled by the Stokes equation, the flow domain $\Omega^{(0)} \subset \mathbb{R}^3$ exhibits a complicated porous geometry. The ability to solve problems on such complicated domains allows applications of numerical upscaling [9]. On the internal boundaries, i. e. on the surface of the grains, no-slip boundary conditions are applied. On the boundaries of the macroscopic domain, we consider Dirichlet and natural Neumann boundary conditions for the velocity. With viscosity μ, velocity $\mathbf{v}$ and pressure p, the problem reads:

$$\begin{aligned}
-\mu\Delta\mathbf{v} + \nabla p &= \mathbf{f} && \text{in} \quad \Omega^{(0)} \\
\nabla\cdot\mathbf{v} &= 0 && \text{in} \quad \Omega^{(0)} \\
\mathbf{v} &= 0 && \text{on} \quad \Gamma_0 \subseteq \partial\Omega^{(0)} \\
\partial_n\mathbf{v} + (p_0 - p)\mathbf{n} &= 0 && \text{on} \quad \Gamma_P = \partial\Omega^{(0)} \setminus \Gamma_0 ,
\end{aligned} \tag{4}$$

For the discretization of the Stokes equations, we use the Discontinuous Galerkin formulation described in [18] which guarantees local mass conservation as well as conservation of momentum. Pressure boundary conditions are imposed as described in [14]. The Discontinuous Galerkin discretization of the Stokes equations reads:

Find $\mathbf{v} \in [V_k^{(i)}]^3$, $p \in V_{k-1}^{(i)}$ such that

$$\begin{aligned}
\mu\, a(\mathbf{v},\mathbf{z}) + b(\mathbf{z},p) &= -\int_{\Gamma_P} p_0\, \mathbf{z}\cdot\mathbf{n}\, ds && \forall\, \mathbf{z} \in [V_k^{(i)}]^3, \\
b(\mathbf{v},q) &= 0 && \forall\, q \in V_{k-1}^{(i)}.
\end{aligned} \tag{5}$$

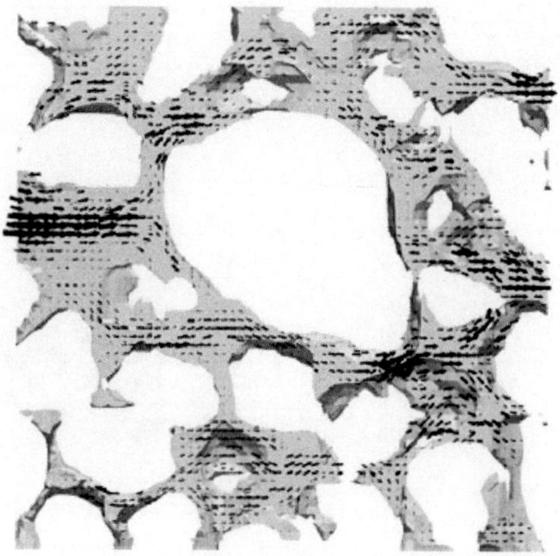

Fig. 3 Slice of the microscopic velocity field through coarse sand. Selection of size 3.6^3mm with a resolution of 64^3 voxels. Original data had a resolution of 256^3 voxels. Pressure drop from left to right and natural boundary conditions on the internal boundaries. Computations are done on a 32^3 mesh with second order ansatz functions for the velocity and first order for the pressure.

where

$$a(\mathbf{v},\mathbf{z}) = \int_\Omega \nabla\mathbf{v} : \nabla\mathbf{z} - \int_\Gamma \{\nabla\cdot\mathbf{v}\}[\![\mathbf{z}]\!] + \int_\Gamma \{\nabla\cdot\mathbf{z}\}[\![\mathbf{v}]\!] \tag{6}$$

$$b(\mathbf{v},q) = -\int_\Omega q\,\nabla\cdot\mathbf{v} + \int_\Gamma \{q\}[\![\mathbf{v}]\!] \tag{7}$$

and where $[\![\cdot]\!]$ and $\{\cdot\}$ denote the component-wise jump and average defined in Section 2.1 and Γ denotes the skeleton of $\boldsymbol{\Omega}^{(0)}$.

The following results were first presented in [9]. By courtesy of H.-J. Vogel we are able to perform computations on experimentally determined pore structures of real soil samples, see Fig. 3. The micro-CT measurements were done at Department Bodenphysik, Helmholtz Zentrum für Umweltforschung (UFZ Halle). The sample is a cylinder filled with coarse sand. It has a diameter of 10mm and a height of 6mm. The data is given as 16bit gray scale values on a $800 \times 828 \times 426$ grid and handled by the `Dune::UDG::MarchingCubeSubTriangulation`.

Numerical upscaling of the permeability for different subsets of the sample yield values in the range between 10^{-10} and $10^{-9}\mathrm{m}^2$ which is in accordance with the values mentioned in the literature [4].

4.2 Incompressible Flow of Two Immiscible Fluids

Let $\boldsymbol{\Omega}$ be an open domain filled with two immiscible fluids. In the following, the index $i = 0,1$ will always refer to the two fluids filling the transient domains $\boldsymbol{\Omega}^{(i)}(t)$ such that $\boldsymbol{\Omega} = \boldsymbol{\Omega}^{(0)}(t) \cup \boldsymbol{\Omega}^{(1)}(t)$ and $\boldsymbol{\Omega}^{(0)}(t) \cap \boldsymbol{\Omega}^{(1)}(t) = \emptyset$ at all times $t \in [0,T)$. The common physical model for such a system is given by:

Let $\mathbf{v} \in C^0(0,T;\boldsymbol{\Omega})$ *with* $\mathbf{v}\big|_{\boldsymbol{\Omega}^{(i)}} \in C^2(0,T;\boldsymbol{\Omega}^{(i)})$ *be a velocity field and* p *with* $p\big|_{\boldsymbol{\Omega}_i} \in C^1(0,T;\boldsymbol{\Omega}^{(i)})$ *a pressure field such that for* $t \in [0,T)$ *and* $\mathbf{v}_0 \in C^2(\boldsymbol{\Omega})$, $\mathbf{v}_D \in C^2(0,T;\boldsymbol{\Omega})$*:*

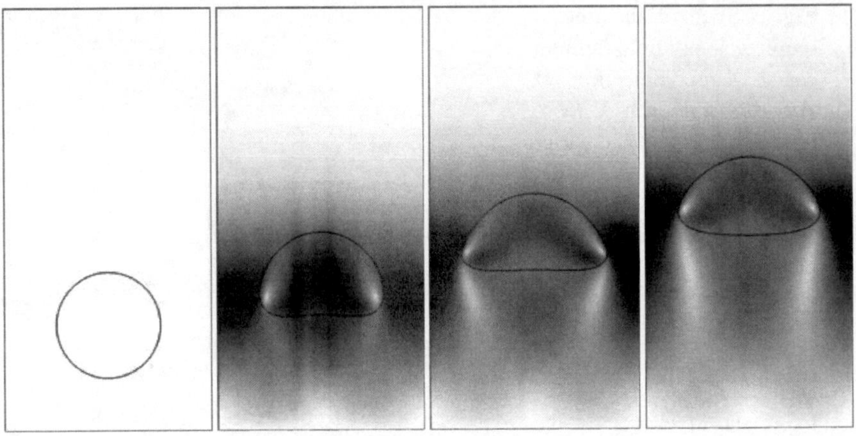

Fig. 4 Buoyancy rise of an initially stationary bubble in a tube with slip boundary conditions. The setup corresponds to a Reynolds number $Re = 35$, an Eotvos number $Eo = 10$, $\rho_0/\rho_1 = 10$, and $\mu_0/\mu_1 = 10$. The grayscale indicates the velocity magnitude.

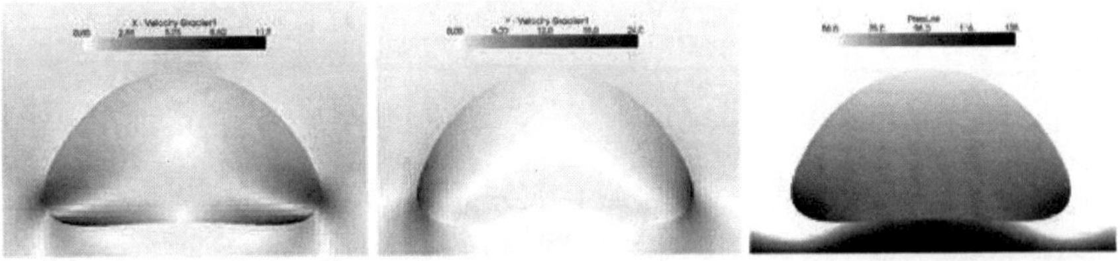

Fig. 5 The pictures show the magnitude of the x-velocity gradients (left), the y-velocity gradients (right) and the pressure field around the bubble during the rise illustrated in Figure 4.

$$
\begin{aligned}
\rho_i \partial_t \mathbf{v} + \rho_i \mathbf{v} \cdot \nabla \mathbf{v} &= -\nabla p + \mu_i \Delta \mathbf{v} && \text{in} \quad \Omega^{(i)}, \\
\nabla \cdot \mathbf{v} &= 0 && \text{in} \quad \Omega, \\
\mathbf{v} &= \mathbf{v}_D && \text{on} \quad \partial\Omega, \\
\mathbf{v}(t=0) &= \mathbf{v}_0 && \text{in} \quad \Omega.
\end{aligned} \tag{8}
$$

Most discretization approaches employ a regularization of the discontinuities in the fluid densities ρ_i and viscosities μ_i which occur at the subdomain boundaries. However, the UDG method allows a direct discretization of the multi-domain formulation as given in (8). Furthermore, at the two fluids' interface $\Gamma = \overline{\Omega}^{(0)} \cap \overline{\Omega}^{(1)}$ with normal vector $\mathbf{n}$ the discontinuous viscosity will result in specific interface conditions:

$$
\forall \mathbf{x} \in \Gamma(t): \qquad [\![\mu(\nabla \mathbf{u} + (\nabla \mathbf{u})^T) - p\mathbf{I}]\!] \mathbf{n}_\Gamma = \mathbf{f}, \tag{9}
$$

These conditions allow for an additional interfacial force $\mathbf{f}$. In the applications presented in [13], it was always chosen to represent a surface tension force according to $\mathbf{f} = \kappa \sigma_s \mathbf{n}$ depending on the surface energy σ_s and the interface curvature κ.

The DG discretization utilized for the single phase flow problem in Section 4.1 may be modified to comply with conditions (9) [13]. Due to the discontinuous basis functions, a highly accurate representation of the discontinuities of the pressure field and the velocity derivatives across Γ may be achieved. This is illustrated in Figure 5.

The temporal evolution of the fluid domains is modeled via the propagation of a level set function ϕ according to

$$\Gamma = \{x \mid \phi(x,t) = 0\} \quad \text{and} \quad \partial_t \phi + \mathbf{v} \cdot \nabla \phi = 0. \tag{10}$$

This scalar function may be used directly for the computation of the sub triangulation in the `Dune::UDG::MarchingCubeSubTriangulation` implementation. For the numerical solution, an explicit Runge-Kutta DG scheme [6] is applied to (10). As we have $\overline{\Omega}^{(0)} \cup \overline{\Omega}^{(1)} = \overline{\Omega}$, the basis for the level set function may be chosen with regard to the cells of the fundamental mesh. Due to the dependence of the interfacial force $\mathbf{f}$ on the interface curvature there are stability requirements connected to the regularity of ϕ in the vicinity of its zero level set. These issues may be addressed by a post-processing of the level set function between time steps. Please refer to [13] for an extensive discussion of these techniques.

Figure 4 shows snapshots of a two-dimensional example application simulating the buoyancy rise of a bubble. The setup corresponds to a benchmark problem as presented in [15] and the simulation results were evaluated quantitatively in [13] to verify and compare the performance of our method.

References

1. Barrett, J.W., Elliott, C.M.: Fitted and unfitted finite-element methods for elliptic equations with smooth interfaces. IMA Journal of Numerical Analysis 7(3), 283–300 (1987)
2. Bastian, P., Blatt, M., Dedner, A., Engwer, C., Klöfkorn, R., Kornhuber, R., Ohlberger, M., Sander, O.: A generic grid interface for parallel and adaptive scientific computing. Part II: Implementation and tests in DUNE. Computing 82(2-3), 121–138 (2008)
3. Bastian, P., Engwer, C.: An unfitted finite element method using discontinuous Galerkin. International Journal for Numerical Methods in Engineering 79(12) (2009)
4. Bear, J.: Dynamics of fluids in porous media. Dover Publications (1988)
5. Belytschko, T., Moes, N., Usui, S., Parimi, C.: Arbitrary discontinuities in finite elements. International Journal for Numerical Methods in Engineering 50(4) (2001)
6. Cockburn, B., Shu, C.W.: The runge-kutta discontinuous Galerkin method for conservation laws v: Multidimensional systems. Journal of Computational Physics 141(2), 199–224 (1998)
7. Coirier, W.J., Powell, K.G.: An accuracy assessment of cartesian-mesh approaches for the euler equations. Journal of Computational Physics 117(1), 121–131 (1995)
8. Dolbow, J.E.: An extended finite element method with discontinuous enrichment for applied mechanics. PhD thesis, Northwestern University (1999)
9. Engwer, C.: An Unfitted Discontinuous Galerkin Scheme for Micro-scale Simulations and Numerical Upscaling. PhD thesis, Universität Heidelberg (2009)

10. Engwer, C., Bastian, P., Kuttanikkad, S.P.: An unfitted discontinuous galerkin finite element method for pore scale simulations. In: 9th International Workshop on State-of-the-Art in Scientific and Parallel Computing. LNCS, Springer, Heidelberg (2008); (in press)
11. Bastian, P., Engwer, C., Fahlke, J., Ippisch, O.: An Unfitted Discontinuous Galerkin method for pore-scale simulations of solute transport. Mathematics and Computers in Simulation 81(10), 2051–2061 (2011), `http://www.sciencedirect.com/science/article/pii/S0378475410004258`
12. Fidkowski, K.J.: A simplex cut-cell adaptive method for high-order discretizations of the compressible Navier-Stokes equations. PhD thesis, Massachusetts Institute of Technology (2007)
13. Heimann, F., Engwer, C., Ippisch, O., Bastian, P.: An unfitted interior penalty discontinuous Galerkin method for incompressible Navier-Stokes two-phase flow. International Journal for Numerical Methods in Fluids (2012), `http://dx.doi.org/10.1002/fld.3653`
14. Heywood, J., Rannacher, R., Turek, S.: Artificial boundaries and flux and pressure conditions for the incompressible Navier–Stokes equations. International Journal for Numerical Methods in Fluids 22, 325–352 (1996)
15. Hysing, S., Turek, S., Kuzmin, D., Parolini, N., Burman, E., Ganesan, S., Tobiska, L.: Quantitative benchmark computations of two-dimensional bubble dynamics. International Journal for Numerical Methods in Fluids 60(11), 1259–1288 (2009)
16. Johansen, H., Colella, P.: A cartesian mesh embedded boundary method for poisson's equation on irregular domains. Journal of Computational Physics 147, 60–85 (1998)
17. Liehr, F., Preusser, T., Rumpf, M., Sauter, S., Schwen, L.O.: Composite finite elements for 3D image based computing. Computing and Visualization in Science 12(4), 171–188 (2009)
18. Rivière, B., Girault, V.: Discontinuous finite element methods for incompressible flows on subdomains with non-matching interfaces. Computer Methods in Applied Mechanics and Engineering 195, 3274–3292 (2006)

Solving Optimal Control Problems with the KASKADE 7 Finite Element Toolbox

Sebastian Götschel, Martin Weiser, and Anton Schiela

Abstract. This paper presents concepts and implementation of the finite element toolbox KASKADE 7, a flexible C++ code for solving elliptic and parabolic PDE systems, based on the DUNE libraries. Issues such as problem formulation, assembly and adaptivity are discussed at the example of optimal control problems. Trajectory compression for parabolic optimization problems is considered as a case study.

1 Introduction

KASKADE 7 is a general-purpose finite element toolbox for solving systems of elliptic and parabolic PDEs. Design targets for the KASKADE 7 code have been *flexibility*, *efficiency*, and *correctness*. One possibility to achieve these, to some extent competing, goals is to use C++ with a great deal of template meta-programming [14]. This generative programming technique uses the C++ template system to let the compiler perform code generation. The resulting code is, due to static polymorphism, at the same time type and const correct and, due to code generation, adapted to the problem to be solved. Since all information relevant to code optimization is directly available to the compiler, the resulting code is highly efficient, of course depending on the capabilities of the compiler. In contrast to explicit code generation, as used, e.g., by the FENICS project [10], no external toolchain besides the C++ compiler/linker is required. Drawbacks of the template meta-programming approach are longer compile times, somewhat cumbersome template notation, and hard to digest compiler diagnostics. Therefore, code on higher abstraction levels, where the performance gains of inlining and avoiding virtual function calls are negligible, uses dynamic polymorphism as well.

The KASKADE 7 code is heavily based on the DUNE libraries [2, 1, 3], which are used in particular for grid management, numerical quadrature, and linear algebra.

Sebastian Götschel · Martin Weiser · Anton Schiela
Zuse Institute Berlin, Takustr. 7, D-14195 Berlin
e-mail: {goetschel,weiser,schiela}@zib.de

As a guiding example at which to illustrate features of KASKADE 7 we will use the all-at-once approach to the following simple optimal control problem. For a desired state y_d defined over a domain $\Omega \subset \mathbb{R}^d$, $d \in \{1,2,3\}$, and $\alpha > 0$ we consider the tracking type problem

$$\min_{u\in L^2(\Omega), y\in H_0^1(\Omega)} \frac{1}{2}\|y-y_d\|_{L^2(\Omega)}^2 + \frac{\alpha}{2}\|u\|_{L^2(\Omega)}^2 \quad \text{s.t.} \quad -\Delta y = u \quad \text{in } \Omega.$$

The solution is characterized by the Lagrange multiplier $\lambda \in H_0^1(\Omega)$ satisfying the Karush-Kuhn-Tucker system

$$\begin{bmatrix} I & & \Delta \\ & \alpha I & I \\ \Delta & I & \end{bmatrix} \begin{bmatrix} y \\ u \\ \lambda \end{bmatrix} = \begin{bmatrix} y_d \\ 0 \\ 0 \end{bmatrix}.$$

For illustration purposes, we will discretize the system using piecewise polynomial finite elements for y and λ and piecewise constant functions for u, even though this is not the best way to approach this particular type of problems [8, 16].

2 KASKADE 7 Structure and Implementation

The foundation of all finite element computation is the approximation of solutions in finite dimensional function spaces. In the next section, we will discuss the representation of functions in KASKADE 7 before addressing the problem formulation.

2.1 *Finite Element Spaces*

On each reference element T_0 there is a set of possibly vector-valued shape functions $\phi_i : T_0 \to \mathbb{R}^s$, $i = 1,\dots,m$ defined. Finite element functions are built from these shape functions by linear combination and transformation. More precisely, finite element functions defined by their coefficient vectors $a \in \mathbb{R}^N$ are given as

$$u(x)|_T = \psi_T(x)(\Phi(\xi)K_T a_{I_T}),$$

where $a_{I_T} \in \mathbb{R}^l$ is the subvector of a containing the coefficients of all finite element ansatz functions which do not vanish on the element T, $K \in \mathbb{R}^{m\times l}$ is a matrix describing the linear combination of shape functions ϕ_i to ansatz functions φ_j, $\Phi(\xi) \in \mathbb{R}^{s\times m}$ is the matrix consisting of the shape functions' values at the reference coordinate ξ corresponding to the global coordinate x as columns, and $\psi_T(x) \in \mathbb{R}^{s\times s}$ is a linear transformation from the values on the reference element to the actual element T.

The indices I_T and efficient application of the matrices K_T and $\psi_T(x)$ are provided by local-to-global-mappers, in terms of which the finite element spaces are defined. The mappers do also provide references to the suitable shape function set, which is, however, defined independently. For the computation of the index set I_T the mappers rely on the DUNE index sets provided by the grid views on which the function spaces are defined.

For Lagrange ansatz functions, the combiner K is just a permutation matrix, and the converter $\psi(x)$ is just 1. For hierarchical ansatz functions in 2D and 3D, nontrivial linear combinations of shape functions are necessary. The implemented over-complete hierarchical FE spaces require just signed permutation matrices [17]. Vectorial ansatz functions, e.g. edge elements, require nontrivial converters $\psi(x)$ depending on the transformation from reference element to actual element. The structure in principle allows to use heterogeneous meshes with different element topology, but the currently implemented mappers require homogeneous meshes of either simplicial or quadrilateral type.

In KASKADE 7, finite element spaces are template classes parameterized with a mapper, defining the type of corresponding finite element functions and supporting their evaluation as well as prolongation during grid refinement, see Sec. 2.4. Assuming that `View` is a suitable DUNE grid view type, FE spaces for the guiding example can be defined as:

```
typedef FEFunctionSpace<ContinuousLagrangeMapper<double,View>
    > H1Space;
typedef FEFunctionSpace<DiscontinuousLagrangeMapper<double,
    View> > L2Space;
H1Space h1Space(gridManager,view,order);
L2Space l2Space(gridManager,view,0);
```

Multi-component FE functions are supported, which gives the possibility to have vector-valued variables defined in terms of scalar shape functions. E.g., displacements in elasto-mechanics and temperatures in the heat equation share the same FE space. FE functions as elements of a FE space can be constructed using the type provided by that space:

```
H1Space::Element<1>::type y(h1Space), lambda(h1Space);
L2Space::Element<1>::type u(l2Space);
```

FE functions provide a limited set of linear algebra operations. Having different types for different numbers of components detects the mixing of incompatible operands at compile time.

During assembly, the ansatz functions have to be evaluated repeatedly. In order not to do this separately for each involved FE function, FE spaces define `Evaluators` doing this once for each involved space. When several FE functions need to be evaluated at a certain point, the evaluator caches the ansatz functions' values and gradients, such that the remaining work is just a small scalar product for each FE function.

2.2 Problem Formulation

For stationary variational problems, the KASKADE 7 core addresses variational functionals of the type

$$\min_{u_i \in V_i} \int_\Omega f(x,u_1,\ldots,u_n,\nabla u_1,\ldots,\nabla u_n)\,dx + \int_{\partial\Omega} g(x,u_1,\ldots,u_n)\,ds \tag{1}$$

The problem definition consists of providing f, g, and their first and second directional derivatives in a certain fashion. First, the number of variables, their number of components, and the FE space they belong to have to be specified. This partially static information is stored in heterogeneous, statically polymorphic containers from the BOOST FUSION [4] library (namespace `bf`). Variable descriptions are parameterized over their space index in the associated container of FE spaces, their number of components, and their unique, contiguous id.

```
typedef bf::vector<H1Space*,L2Space*> Spaces;
Spaces spaces(&h1Space,&l2Space);
typedef bf::vector<VariableDescription<0,1,0>,
    VariableDescription<0,1,1>, VariableDescription<1,1,2> >
    VarDesc;
```

Besides this data, a problem class defines, apart from some static meta information, two mandatory member classes, the `DomainCache` defining f and the `BoundaryCache` defining g. The domain cache provides member functions `d0`, `d1`, and `d2` evaluating $f(\cdot)$, $f'(\cdot)v_i$, and $f''(\cdot)[v_i,w_j]$, respectively. For the guiding example with

$$f = \frac{1}{2}(y-y_d)^2 + \frac{\alpha}{2}u^2 + \nabla\lambda^T\nabla y - \lambda u,$$

the corresponding code looks like

```
double d0() const {
  return (y-yd)*(y-yd)/2 + u*u*alpha/2 + dlambda*dy - lambda*
      u;
}
template <int i, int d>
double d1(VariationalArg<double,d> const& vi) const {
  if (i==0) return (y-yd)*vi.value + dlambda*vi.gradient;
  if (i==1) return alpha*u*vi.value - lambda*vi.value;
  if (i==2) return dy*vi.gradient - u*vi.value;
}
template <int i, int j, int d>
double d2(VariationalArg<double,d> const& vi,
          VariationalArg<double,d> const& wj) const {
  if (i==0 && j==0) return vi.value*wj.value;
  if (i==0 && j==2) return vi.gradient*wj.gradient;
  if (i==1 && j==1) return alpha*vi.value*wj.value;
  if (i==1 && j==2) return -vi.value*wj.value;
  if (i==2 && j==0) return vi.gradient*wj.gradient;
  if (i==2 && j==1) return -vi.value*wj.gradient;
}
```

A static member template class D2 defines which Hessian blocks are available. Symmetry is auto-detected, such that in d2 only $j \leq i$ needs to be defined.

```
template <int row, int col>
class D2 {
  static int present = (row==2) || (row==col);
};
```

The boundary cache is defined analogously.

The functions for y, u, and λ are specified (for nonlinear or instationary problems in form of FE functions) on construction of the caches, and can be evaluated for each quadrature point using the appropriate one among the evaluators provided by the assembler:

```
template <class Pos, class Evaluators>
void evaluateAt(Pos const& x, Evaluators const& evaluators) {
    y = yFunc.value(at_c<0>(evaluators));
    u = uFunc.value(at_c<1>(evaluators));
    lambda = lambdaFunc.value(at_c<0>(evaluators));
    dy = yFunc.gradient(at_c<0>(evaluators));
    dlambda = lambdaFunc.gradient(at_c<0>(evaluators));
}
```

2.3 *Assembly*

Assembly of matrices and right hand sides for variational functionals is provided by the template class `VariationalFunctionalAssembler`, parameterized with a (linearized) variational functional. The elements of the grid are traversed. For each cell, the functional is evaluated at the integration points provided by a suitable quadrature rule, assembling local matrices and right hand sides. If applicable, boundary conditions are integrated. Finally, local data is scattered into global data structures. Matrices are stored as sparse block matrices with compressed row storage, as provided by the DUNE `BCRSMatrix<BlockType>` class. For evaluation of FE functions and management of degrees of freedom, the involved spaces have to be provided to the assembler. User code for assembling a given functional will look like the following:

```
bf::vector<H1Space*,L2Space*> spaces(&h1space, &l2space);
VariationalFunctionalAssembler<Functional> as(spaces);
as.assemble(linearization(f,x));
```

For the solution of the resulting linear systems, several direct and iterative solvers can be used through an interface to DUNE-ISTL. For instance, the DUNE `AssembledLinearOperator` interface is provided by the KASKADE 7 class `AssembledGalerkinOperator`. After the assembly, and some more initializations (`rhs`, `sol`), e.g. a direct solver `solverType` can be applied:

```
AssembledGalerkinOperator A(as);
directInverseOperator(A,solverType).applyscaleadd(-1.,rhs,sol);
```

2.4 Adaptivity

KASKADE 7 provides several means of error estimation.

Embedded error estimator. Given a FE function u, an approximation of the error can be obtained by projecting u onto the ansatz space with polynomials of order one less. The method `embeddedErrorEstimator()` then constructs (scaled) error indicators, marks cells for refinement and adapts the grid with aid of the `GridManager` class, which will be described later.

```
error = u;
projectHierarchically(variableSet, u);
error -= u;
accurate = embeddedErrorEstimator(variableSet,error,u,scaling,
    tol,gridManager);
```

Hierarchic error estimator. After discretization using a FE space S_l, the minimizer of the variational functional satisfies a system of linear equations, $A_{ll}x_l = -b_l$. For error estimation, the ansatz space is extended by a second, higher order ansatz space, $S_l \oplus V_q$. The solution in this enriched space satisfies

$$\begin{bmatrix} A_{ll} & A_{lq} \\ A_{ql} & A_{qq} \end{bmatrix} \begin{bmatrix} x_l \\ x_q \end{bmatrix} = - \begin{bmatrix} b_l \\ b_q \end{bmatrix}.$$

Of course the solution of this system is quite expensive. As x_l is essentially known, just the reduced system $\operatorname{diag}(A_{qq})x_q = -(b_q + A_{ql}x_l)$ is solved [6]. A global error estimate can be obtained by evaluating the scalar product $\langle x_q, b_q \rangle$.

In KASKADE 7, the template class `HierarchicErrorEstimator` is available. It is parameterized by the type of the variational functional, and the description of the hierarchic extension space. The latter can be defined using e.g. the `ContinuousHierarchicExtensionMapper`. The error estimator then can be assembled and solved, analogously to the assembly and solution of the original variational functional.

Grid transfer. Grid transfer makes heavy use of the signal-slot concept, as implemented in the BOOST.SIGNALS library [4]. Signals can be seen as callback functions with multiple targets. They are connected to so-called slots, which are functions to be executed when the signal is sent. This paradigm allows to handle grid modifications automatically, ensuring that all grid functions stay consistent.

All mesh modifications are done via the `GridManager<Grid>` class, which takes ownership of a grid once it is constructed. Before adaptation, the grid manager triggers the affected FE spaces to collect necessary data in a class `TransferData`. For all cells, a local restriction matrix is stored, mapping global degrees of freedom to local shape function coefficients of the respective father cell. After grid refinement or coarsening, the grid manager takes care that all FE functions are transferred to the new mesh. Since the construction of transfer matrices from grid modifications is a computationally complex task, these matrices are constructed only once for each FE space. On that account, FE spaces listen for the `GridManager`'s signals. As soon as the transfer matrices are constructed, the FE spaces emit signals to which the associated FE functions react by updating their coefficient vectors using the provided transfer matrix. Since this is just an efficient linear algebra operation, transferring quite a lot of FE functions from the same FE space is cheap.

After error estimation and marking, the whole transfer process is initiated in the user code by:

```
gridManager.adaptAtOnce();
```

The automatic prolongation of FE functions during grid refinement makes it particularly easy to keep coarser level solutions at hand for evaluation, comparison, and convergence studies.

2.5 Time-Dependent Problems

KASKADE 7 provides an extrapolated linearly implicit Euler method for integration of time-dependent problems $B(y)\dot{y} = f(y)$, [7]. Given an evolution equation `Equation eq`, the corresponding loop looks like

```
Limex<Equation> limex(gridManager,eq,variableSet);
for (steps=0; !done && steps<maxSteps; ++steps) {
  do {
    dx = limex.step(x,dt,extrapolOrder,tolX);
    errors = limex.estimateError(/*...*/);
    // ... (choose optimal time step size)
  } while( error > tolT );
  x += dx ;
}
```

Step computation makes use of the class `SemiImplicitEulerStep`. Here, the stationary elliptic problem resulting from the linearly implicit Euler method is defined. This requires an additional method `b2` in the domain cache for the evaluation of B. For the simple scalar model problem with $B(x)$ independent of y, this is just the following:

```
template<int i, int j, int d>
Dune::FieldMatrix<double, TestVars::Components<i>::m, AnsatzVars
    ::Components<j>::m>
  b2(VariationalArg<double,d> const& vi, VariationalArg<double,d
      > const& wj) const {
    return bvalue*vi.value*vj.value;
  }
```

Of course, `bvalue` has to be specified in the `evaluateAt` method.

2.6 Nonlinear Solvers

A further aspect of KASKADE 7 is the solution of nonlinear problems, involving partial differential equations. Usually, these problems are posed in function spaces, which reflect the underlying analytic structure, and thus algorithms for their solution should be designed to inherit as much as possible from this structure.

Algorithms for the solution of nonlinear problems of the form (1) build upon the components described above, such as discretization, iterative linear solvers, and adaptive grid refinement. A typical example is Newton's method for the solution of a nonlinear operator equation. Algorithmic issues are the adaptive choice of damping factors, and the control of the accuracy of the linear solvers. This includes requirements for iterative solvers, but also requirements on the accuracy of the discretization.

The interface between nonlinear solvers and supporting routines is rather coarse grained, so that dynamic polymorphism is the method of choice. This makes it possible to develop and compile complex algorithms independently of the supporting routines, and to reuse the code for a variety of different problems. In client code the components can then be plugged together, and decisions are made, which type of discretization, linear solver, adaptivity, etc. is used together with the nonlinear algorithm. In this respect, KASKADE 7 provides a couple of standard components, but of course users can write their own specialized components.

Core of the interface are abstract classes for a mathematical vector, which supports vector space operations, but no coordinate-wise access, abstract classes for norms and scalar products, and abstract classes for a nonlinear functional and its linearization (or, more accurately, its local quadratic model). Further, an interface for inexact linear solvers is provided. These concepts form a framework for the construction of iterative algorithms in function space, which use discretization for the computation of inexact steps and adaptivity for error control.

In order to apply an algorithm to the solution of a nonlinear problem, one can in principle derive from these abstract classes and implement their purely virtual methods. However, for the interaction with the other components of KASKADE 7, bridge classes are provided, which are derived from the abstract base classes, and own an implementation.

We explain this at the following example which shows a simple implementation of the damped Newton method:

```
for(step=1; step <= maxSteps; step++) {
  lin = functional->getLinearization(*iterate);
  linearSolver->solve(*correction,*lin);
  do {
    *trialIter = *iterate;
    trialIter->axpy(dampingFactor,*correction);
    if(regularityTest(dampingFactor)==Failed) return -1;
    updateDampingFactor(dampingFactor);
  }
  while(evalTrialIterate(*trialIter,*correction,*lin)==Failed);
  *iterate = *trialIter;
  if(convergenceTest(*correction,*iterate)==Achieved) return 1;
}
```

While `regularityTest`, `updateDampingFactor`, `evalTrialIterate`, and `convergenceTest` are implemented within the algorithm, `functional`, `lin`, and `linearSolver`, used within the subroutines are instantiations of derived classes, provided by client code. By

```
linearSolver->solve(*correction,*lin);
```

a linear solver is called, which has access to the linearization `lin` as a linear operator equation. It may either be a direct or an iterative solver on a fixed discretization, or solve this operator equation adaptively, until a prescribed relative accuracy is reached. In the latter case, the adaptive solver calls in turn a linear solver on each refinement step. There is a broad variety of linear solvers available, and moreover, it is not difficult to implement a specialized linear solver for the problem at hand.

The object `lin` is of type `AbstractLinearization`, which is implemented by the bridge class `Bridge::KaskadeLinearization`. This bridge class is a template, parametrized by a variational functional and a vector of type `VariableSet::Representation`. It uses the assembler class to generate the data needed for step computation and manages generated data. From the client side, only the variational functional has to be defined and an appropriate set of variables has to be given.

Several algorithms are currently implemented. Among them there is a damped Newton method [5] with affine covariant damping strategy, a Newton path-following algorithm, and algorithms for nonlinear optimization, based on a cubic error model. This offers the possibility to solve a large variety of nonlinear problems involving partial differential equations. As an example, optimization problems with partial differential equations subject to state constraints can be solved by an interior point method combining Newton path-following and adaptive grid refinement [13].

3 State Trajectory Compression

As a case study we consider a 3D parabolic model optimal control problem

$$\min \frac{1}{2} \|y - y_d\|^2_{L^2(\Omega\times(0,T))} + \frac{\alpha}{2} \|u\|^2_{L^2(\partial\Omega\times(0,T))},$$

subject to

$$By_t - \Delta y = f \text{ in } \Omega \times (0,T), \quad \partial_\nu y + y = u \text{ on } \partial\Omega \times (0,T), \quad y(\cdot,0) = 0 \text{ in } \Omega.$$

To avoid 4D discretization, methods working on the reduced objective functional are often employed. For the computation of the reduced gradient, one forward solve of the state equation as well as one backward solve of the adjoint equation is needed, see e.g. [8] and the references therein. As the state enters into the adjoint equation, the forward solution, a 4D data set, has to be stored. In this section, we will focus on the implementation of a compression scheme in KASKADE 7.

3.1 A Lossy Compression Algorithm

Assume a nested family $\mathscr{T}_0 \subset \cdots \subset \mathscr{T}_l$ of triangulations, constructed from a coarse grid $\mathscr{T}_0$ by uniform or adaptive refinement. With $\mathscr{N}_j$ denoting the set of nodes on level j, let $\{\varphi_k | k \in \mathscr{N}_j\}$ be the piecewise linear nodal basis functions over the triangulation $\mathscr{T}_j$. Thus, the solution of the state equation is given by $y(x,t_i) = \sum_{k\in\mathscr{N}_l} y_{k,i}\varphi_k(x)$.

The compression relies on a predictor for the nodal values $y_{k,i}$, combined with entropy coding of the prediction errors. As the data is used inside an discretized, iterative algorithm, lossy compression maintaining a certain error bound is sufficient to ensure convergence.

At time t_i we make use of the mesh hierarchy. Initially, the finite element coefficients for the coarse grid are predicted as zero. Given an approximation of $y_{k,i}, k = 0,\ldots,|\mathscr{N}_j| - 1$ on grid level j, a prolongation of these values to the next grid level $j+1$ is computed, e.g. by linear interpolation. As the exact nodal values are known, the prediction error can be evaluated, quantized keeping an error bound, and stored. The reconstruction of the nodal values then is used as input for the next prediction step, allowing the decoder to mirror the prediction process during the adjoint integration.

For 2D problems, uniform refinement and sufficiently smooth states y, a theoretical estimate of the expected performance can be derived. With an L^∞ quantization error not exceeding the discretization error, approximately 2.9 bits/vertex are needed to store the state values at a single timestep, yielding a compression factor of 22 compared to storing floating point values at 64 bits/vertex.

A detailed discussion of the compression scheme, the theoretical derivation and further examples can be found in [15].

3.2 Implementation

Implementation of the lossy compression algorithm makes use of the existing infrastructure for adaptivity, see Sec. 2.4. For the prolongation step, a FE space over the current `LevelGridView` is constructed. With help of the `TransferData` class, a transfer matrix is generated, which applied to a suitable coefficient vector `coeff` performs the transfer to the next grid level. This differs from the usual grid transfer, where the coefficient vector is prolongated to the *leaf* grid. For code re-usability, this difference is taken care of by policy classes `AdaptationCoarseningPolicy` and `MultilevelCoarseningPolicy`, which control the acceptance of cells during the creation of transfer matrices:

```
MultilevelCoarseningPolicy policy(level) ;
Space space(gridManager, grid.levelView(level), order);
TransferData<Space,MultilevelCoarseningPolicy> transferData(
    space, policy);
space.setGridView(grid.levelView(level+1));
newCoeff = transferData.transferMatrix().apply(coeff) ;
```

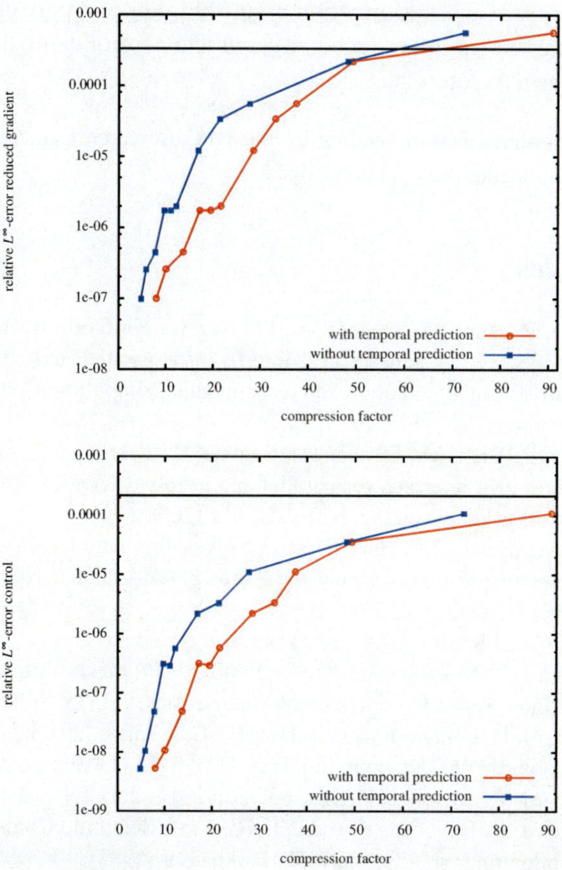

Fig. 1 Relative error vs. compression rate for reduced gradient (top) and control (bottom) after 100 gradient steps. The horizontal line shows the approximated discretization error.

Evaluation of the prediction error is straightforward, as the FE solution on the leaf grid is at hand. For quantization, the range of prediction errors on the current grid level is divided into N subintervals, where $N = \text{range}/(2\delta)$ is determined by the required quantization error tolerance δ. For additional compression in time, differential encoding is used for the subinterval indices, i. e. only the differences between to timesteps are stored using a range coder [11].

For decoding the state values during the solution of the adjoint equation, the corresponding adaptively refined grids have to be stored as well. An efficient algorithm using the hierarchic structure of the mesh can be found in [9].

We apply the compression scheme to the model optimal control problem with data $\alpha = 10^{-5}$, $y_d(x,t) = t|x|^2$, $f(x,t) = |x|^2 - 4t$, discretized in space with linear finite elements on a uniform mesh. For minimization, a simple gradient-descent algorithm with an Armijo stepsize rule is used (see e.g. [12]). Fig. 1 shows the error in the reduced gradient and control induced by the lossy compression scheme, after

100 iterations of the optimization algorithm. For comparison, we estimate the discretization error for the reduced gradient and control using the quantities computed on a finer grid as reference.

Acknowledgment. Partial funding by the DFG Research Center MATHEON, projects A1, A17, and F9, is gratefully acknowledged.

References

1. Bastian, P., Blatt, M., Dedner, A., Engwer, C., Klöfkorn, R., Kornhuber, R., Ohlberger, M., Sander, O.: A generic grid interface for parallel and adaptive scientific computing. Part II: Implementation and tests in dune. computing. Computing 82(2-3), 121–138 (2008)
2. Bastian, P., Blatt, M., Dedner, A., Engwer, C., Klöfkorn, R., Ohlberger, M., Sander, O.: A generic grid interface for parallel and adaptive scientific computing. Part I: Abstract framework. Computing 82(2-3), 103–119 (2008)
3. Blatt, M., Bastian, P.: The Iterative Solver Template Library. In: Kågström, B., Elmroth, E., Dongarra, J., Waśniewski, J. (eds.) PARA 2006. LNCS, vol. 4699, pp. 666–675. Springer, Heidelberg (2007)
4. Boost: C++ libraries, `http://www.boost.org/`
5. Deuflhard, P.: Newton Methods for Nonlinear Problems. Affine Invariance and Adaptive Algorithms. Series Computational Mathematics, vol. 35. Springer (2006)
6. Deuflhard, P., Leinen, P., Yserentant, H.: Concepts of an adaptive hierarchical finite element code. IMPACT Comp. Sci. Eng. 1(1), 3–35 (1989)
7. Deuflhard, P., Nowak, U.: Extrapolation integrators for quasilinear implicit ODEs. In: Deuflhard, P., Engquist, B. (eds.) Large Scale Scientific Computing, Progress in Scientific Computing, vol. 7, pp. 37–50, Birkhäuser (1987)
8. Hinze, M., Pinnau, R., Ulbrich, M., Ulbrich, S.: Optimization with PDE constraints. Springer, Berlin (2009)
9. Kälberer, F., Polthier, K., von Tycowicz, C.: Lossless compression of adaptive multiresolution meshes. In: Proc. Brazilian Symposium on Computer Graphics and Image Processing (SIBGRAPI), vol. 22 (2009)
10. Logg, A.: Automating the finite element method. Arch. Comput. Methods Eng. 14, 93–138 (2007)
11. Martin, G.: Range encoding: an algorithm for removing redundancy from a digitised message. Presented at Video & Data Recording Conference, Southampton (1979)
12. Nocedal, J., Wright, S.J.: Numerical Optimization. Springer, New York (2006)
13. Schiela, A., Günther, A.: An interior point algorithm with inexact step computation in function space for state constrained optimal control. Numer. Math. 119(2), 373–407 (2011)
14. Veldhuizen, T.: Using C++ template metaprograms. C++ Report 7(4), 36–43 (1995)
15. Weiser, M., Götschel, S.: State trajectory compression for optimal control with parabolic PDEs. SIAM J. Sci. Comput. 34(1), A161–A184 (2012)
16. Weiser, M., Gänzler, T., Schiela, A.: Control reduced primal interior point methods. Comput. Optim. Appl. 41(1), 127–145 (2008)
17. Zumbusch, G.: Symmetric hierarchical polynomials and the adaptive h-p-version. In: Ilin, A., Scott, L. (eds.) Proc. of the Third Int. Conf. on Spectral and High Order Methods, ICOSAHOM 1995. Houston Journal of Mathematics, pp. 529–540 (1996)

Part III
Specific Methods and Applications

The DuMu^x Material Law Framework

Andreas Lauser

Abstract. Numerical simulation of multi-phase, multi-component flow in porous media is a very challenging task. Besides conceptional issues like the proper separation of scales, this is primarily because of the strong non-linearities caused by the heterogeneities of the porous medium as well as the complicated thermodynamics of miscible multiphase flow. This document tries to first provide a rough overview on the relevant parts of thermodynamics for multiple fluid phases and multiple chemical species in a porous medium. It then proceeds to the implementation of this abstract framework in the DUNE based numerical simulation toolbox DuMu^x which is, like DUNE, publicly available as free software under the terms of the General Public License (GPL).

1 Introduction

Fluid flow and transport processes in porous media are relevant for a wide range of technical, environmental and biological systems. As a consequence, there are many practical applications where flow processes through porous media are relevant; for example polymer electrolyte membrane fuel cells, ground water extraction, soil remediation, oil extraction and cancer therapy.

In this context problems which involve multiple miscible fluid phases and multiple chemical species are especially difficult to handle numerically. Such applications are, for example, oil extraction and steam assisted soil remediation.

Often, numerical simulation frameworks for flow and transport in porous media implicitly make thermodynamic assumptions and material laws are also often implemented in an ad-hoc manner. This easily leads to thermodynamic inconsistencies which limit the predictiveness of the simulations and often imply sub-optimal nu-

Andreas Lauser
Department of Hydrodynamics and Modeling of Hydrosystems, University of Stuttgart
e-mail: Andreas.Lauser@iws.uni-stuttgart.de

merical performance. Thus, before outlining implementation details, this paper first gives an abstract description of the thermodynamics involved in multi-phase flows through porous media, with the simplifying assumptions being explicitly mentioned.

This work is structured as follows: First, a very brief introduction to multi-phase flow in porous media will be given and the differential equations for mass balance will be motivated in section 2. Because the differential equations alone do not define a closed system of equations, closure conditions must be applied. What can be used is outlined in section 3. In this section, a special focus lies on local thermodynamic equilibrium. Before concluding in section 5, section 4 sketches the implementation of the theoretical frameworks outlined in sections 2 and 3 in the DUNE based flow simulator DuMu^x.

2 Multi-phase, Multi-component Flow and Transport in Porous Media

The most important equations of multi-phase, multi-component flow and transport processes in porous media are the equations for the conservation of mass. For each chemical species κ, this mass balance is captured by the partial differential equation

$$\sum_\alpha \left(\frac{\partial(\phi\, S_\alpha\, x_\alpha^\kappa / V_{m\alpha})}{\partial t} + \mathrm{div}\left\{ \frac{x_\alpha^\kappa}{V_{m\alpha}}\, \mathbf{v}_\alpha \right\} \right) = 0\,, \tag{1}$$

where α stands for a *fluid phase* (often either water, oil or gas), ϕ for the *porosity* (ratio of pore volume to the total volume), S_α is the *saturation* of the fluid α (amount of pore volume filled by phase α), $V_{m\alpha}$ is the *molar volume* of phase α (volume occupied by one mole of the phase), x_α^κ is the *mole fraction* of the component κ in phase α (ratio of moles of chemical species κ to the total number of moles in the phase) and $\mathbf{v}_\alpha$ is the *filter velocity* of phase α.

In general, the velocities $\mathbf{v}_\alpha$ need to be determined implicitly using momentum balance equations; for example, using the Navier-Stokes equations which apply for Newtonian fluids. However, in the context of porous media, Darcy's law is usually assumed to be valid. For multi-phase flow, this relation is expressed by

$$\mathbf{v}_\alpha = -\frac{k_{r\alpha}}{\mu_\alpha} \mathbf{K}\,\mathbf{grad}\left[p_\alpha - g \frac{\overline{M}_\alpha}{V_{m\alpha}} \right]\,, \tag{2}$$

where $\mathbf{K}$ expresses the *intrinsic permeability* of the porous medium by means of a symmetric, positive definite tensor, μ_α is the *dynamic viscosity* of the fluid, p_α is the fluid *pressure*, g is the *gravity potential*, $\overline{M}_\alpha$ is the *mean molar mass* and $k_{r\alpha}$ is the *relative permeability*. The relative permeability, is in the range $[0,1]$ and models the fact that if more than one fluid phase is involved, some of the medium's pores are occupied by other phases and the filter velocity of α is thus reduced. Since (2) is explicit in $\mathbf{v}_\alpha$, it can be directly inserted back into (1). This yields the governing partial differential equations for the conservation of component masses within a multi-phase, multi-component flow in porous media [5],

$$\sum_\alpha \left(\frac{\partial(\phi\, S_\alpha\, x_\alpha^\kappa / V_{m\alpha})}{\partial t} - \operatorname{div}\left\{ \frac{x_\alpha^\kappa}{V_{m\alpha}} \frac{k_{r\alpha}}{\mu_\alpha} \mathbf{K}\, \mathbf{grad} \left[p_\alpha - g \frac{\overline{M}_\alpha}{V_{m\alpha}} \right] \right\} \right) = 0. \quad (3)$$

Also, (3) can be extended to include effects like dispersion or molecular diffusion [2] and conservation of energy can also be considered in addition to the mass balances; Although, for the sake of simplicity, these are not outlined here.

3 Closure Conditions

If M fluid phases and N chemical species are involved, (3) defines N equations, but the total number of quantities occurring in (3) is $M \cdot (N+6) + 3$. These are the porosity ϕ and intrinsic permeability of the porous medium $\mathbf{K}$, the gravity potential g, and, for each of the M fluid phases, pressure p_α, saturation S_α, relative permeability $k_{r\alpha}$, dynamic viscosity μ_α, mean molar mass $\overline{M}_\alpha$ and composition $\{x_\alpha^1, \dots, x_\alpha^N\}$. Of those quantities, $\mathbf{K}$, g and ϕ are externally specified parameters, which reduces the number of degrees of freedom which need to be determined by constitutive relations to $M \cdot (N+6) - N$. The following subsections will outline how these constitutive relations can be obtained.

3.1 *Saturation Closure Condition*

One additional equation can immediately obtained from the fact that all pores must be filled by some fluid: This implies that the sum of the saturations of all phases is equal to 100%, i.e.

$$\sum_\alpha S_\alpha = 1 \qquad \text{holds.} \quad (4)$$

3.2 *Mean Molar Mass*

The mean molar mass $\overline{M}_\alpha$ any phase α is also easily defined as the sum of the constituents' molar masses weighted by their respective mole fraction:

$$\overline{M}_\alpha = \sum_\kappa M^\kappa x_\alpha^\kappa . \quad (5)$$

Table 1 Parameters for the common cubic thermal equations of state as presented in [8]. T^{κ}_{crit} and p^{κ}_{crit} correspond to the critical temperature and critical pressure of a pure component κ, and ω^{κ} is the acentric factor of the component. All these quantities can either be taken from the literature, or can be estimated, see [8].

Name	u_α	w_α	b^{κ}_{α}	a^{κ}_{α}
van der Waals	0	0	$\frac{RT^{\kappa}_{crit}}{8p^{\kappa}_{crit}}$	$\frac{27}{64}\frac{(RT^{\kappa}_{crit})^2}{p^{\kappa}_{crit}}$
Redlich-Kwong	1	0	$\frac{0.08664RT^{\kappa}_{crit}}{p^{\kappa}_{crit}}$	$\frac{0.42748R^2T^{\kappa}_{crit})^{5/2}}{p^{\kappa}_{crit}\sqrt{T}}$
Soave	1	0	$\frac{0.08664RT^{\kappa}_{crit}}{p^{\kappa}_{crit}}$	$\frac{0.42748(RT^{\kappa}_{crit})^2}{p^{\kappa}_{crit}}\left[1+(1-\sqrt{T_r})\,f_{\omega^{\kappa}}\right]^2$ with $T_r = T_\alpha/T^{\kappa}_{crit}$ and $f_{\omega^{\kappa}} = 0.48+\omega^{\kappa}(1.574-0.176\omega^{\kappa})$
Peng-Robinson	2	-1	$\frac{0.07780RT^{\kappa}_{crit}}{p^{\kappa}_{crit}}$	$\frac{0.45724(RT^{\kappa}_{crit})^2}{p^{\kappa}_{crit}}\left[1+(1-\sqrt{T_r})\,f_{\omega^{\kappa}}\right]^2$ where $f_{\omega^{\kappa}} = 0.37464+\omega^{\kappa}(1.54226-0.26992\omega^{\kappa})$

3.3 *Thermal Equations of State*

Next, a thermal equation of state needs to be chosen for each phase α. The thermal equation of state describes the relation between the molar volume $V_{m\alpha}$, phase pressure p_α, *phase temperature* T_α, and phase composition $\{x^1_\alpha,\dots,x^N_\alpha\}$:

$$p_\alpha = p_\alpha(V_{m\alpha},T_\alpha,x^1_\alpha,\dots,x^N_\alpha)\,. \tag{6}$$

Cubic Equations of State

The most commonly used equations of state can be expressed as cubic rational functions in terms of the molar volume $V_{m\alpha}$ with the defining equation

$$p_\alpha = \frac{RT_\alpha}{V_{m\alpha}-b_\alpha} - \frac{a_\alpha}{V^2_{m\alpha}+u_\alpha b_\alpha V_{m\alpha}+w_\alpha(b_\alpha)^2}\,. \tag{7}$$

Here, R is the *ideal gas constant*, u_α and v_α are approach specific constants, b_α can be interpreted as the volume occupied by one mole of molecules of the phase (often referred to as the *covolume*) and a_α is the *attractive factor* for the phase α.

For fluid mixtures, the parameters a_α and b_α are determined from the parameters of the pure components by means of a mixing rule. For cubic equations of state, the relations

$$a_\alpha = \sum_\kappa\sum_\lambda x^\kappa_\alpha x^\lambda_\alpha\sqrt{a^\kappa_\alpha a^\lambda_\alpha}(1-\bar{k}^{\kappa\lambda}_\alpha) \qquad \text{and} \tag{8}$$
$$b_\alpha = \sum_\kappa x^\kappa_\alpha b^\kappa_\alpha\,,$$

Fig. 1 A simple porous medium with the three fluid phases: A liquid primarily composed of polar molecules (water), a liquid primarily consisting of apolar molecules (oil) and gas. All three fluid phases are mixtures of the same components with unequal compositions. These components might be, for example, H_2O, CH_4 and C_2HCl_3 (trichloroethylene). Due to the capillary effect, the phases exhibit distinct pressures even under stationary conditions.

can be used [8]. Here, $\bar{k}_{\alpha}^{\kappa\lambda}$ is the *interaction coefficient* between components κ and λ – which can be found in the literature – and $a_{\alpha}^{\langle\cdot\rangle}, b_{\alpha}^{\langle\cdot\rangle}$ are the parameters for the pure components as given by Table 1.

3.4 Local Thermodynamic Equilibrium

Most models used to simulate flow through porous media assume local thermodynamic equilibrium. In this context, *local* means, that any infinitesimally small sub-domain of the simulated domain which is insulated from the rest of the system at any time, is at thermodynamic equilibrium.

Thermodynamic equilibrium in the porous media context can be thought of as the experiment depicted in Figure 1 at stationary conditions. It is thus the superposition of the following three sub-equilibria:

- **Thermal Equilibrium**, defined as the temperature of all phases being constant with time,
- **Mechanical Equilibrium**, meaning that the pressure of all phases is constant in time and
- **Chemical Equilibrium**, which means that the composition of any phase does not change with time.

In the following, the three equilibria will be discussed independently of each other.

Thermal Equilibrium

Thermal equilibrium implies that the temperature in all fluid phases is equal, i.e.

$$T_{\beta} = T_{\gamma} =: T \qquad \text{for any phases } \beta, \gamma. \tag{9}$$

If energy is conserved using an partial differential equation in addition to (3), this suggests to use the temperature T as a primary variable, otherwise T must be given externally.

Mechanical Equilibrium

The consequence of mechanical equilibrium is that the pressure between fluid phases is stationary, i.e. that the difference between any fluid phases β and γ can be described using a closed function

$$p_{c\beta\gamma} := p_\beta - p_\gamma \,. \tag{10}$$

This function is usually called the *capillary pressure* between phases β and γ. In the most general case, $p_{c\beta\gamma}$ is a function of all fluid saturations $\{S_\alpha\}$ for $\alpha \in \{1,\ldots,M\}$, all phase compositions $\{x_\alpha^\kappa\}$, phase temperatures $\{T_\alpha\}$, as well as the actual pore scale distribution of the phases. Similarly, the relative permeability of a phase $k_{r\alpha}$ in general is also a function of those parameters. Common relations for capillary pressure and relative permeability like the Brooks-Corey [1] and van Genuchten [10] models assume that $p_{c\beta\gamma}$ is only a function of the saturations $\{S_\alpha\}$, though.

Chemical Equilibrium

Stationary composition of the phases in a multi-phase, multi-component context means that no net-mass of any component is transferred from one phase to another. Thermodynamically this is equivalent to the statement that the system cannot further decrease its internal energy by changing the composition of fluid phases relative to each other. This means that the *chemical potential* of each component is the same in all phases

$$\forall\kappa\forall\alpha,\beta\colon\ \zeta_\alpha^\kappa = \zeta_\beta^\kappa =: \zeta^\kappa \,, \tag{11}$$

where ζ^κ denotes the chemical potential of a component κ[1]. Note that ζ^κ is well-defined if chemical equilibrium is assumed.

Since chemical potentials become infinite for components with a mole fraction of zero, they are numerically impractical. Instead, *fugacities* $f_{(\cdot)}^\kappa$ are usually used. Fugacity is a quantity with SI-unit $[Pa]$ that can be interpreted as "urgency of a component κ to leave a fluid phase α". It is related to the chemical potential by the relation [7]

$$f_\alpha^\kappa = \exp\left\{\frac{\zeta_\alpha^\kappa}{k_B T_\alpha}\right\} \,, \tag{12}$$

[1] In the thermodynamic literature the symbol for the chemical potential usually is μ. In order to avoid confusion with the the dynamic viscosity, ζ is used here, though.

where $k_B = 1.3806 \cdot 10^{-23}\, J/K$ is the Boltzmann constant. Since (12) is an injective function, it implies in conjunction with (11) and (9) that for any two phases β and γ

$$f^\kappa_\beta = f^\kappa_\gamma =: f^\kappa$$

holds for any component κ.

Usually, the fugacity of a component in a phase is calculated via a so-called fugacity coefficient Φ^κ_α which is defined via

$$f^\kappa_\alpha = x^\kappa_\alpha p_\alpha \Phi^\kappa_\alpha \,. \tag{13}$$

For many equations of state, there are closed formulas for Φ^κ_α; for the cubic equations of state outlined in Section 3.3, the fugacity of a component κ in a fluid phase α is given by the expression [8]

$$\begin{aligned}\ln \Phi^\kappa_\alpha =& \frac{b^\kappa_\alpha}{b}(Z_\alpha - 1) - \ln(Z_\alpha - B^{\star\kappa}_\alpha) \\ &+ \frac{A^{\star\kappa}_\alpha}{B^\star \sqrt{u^2_\alpha - 4w_\alpha}} \left(\frac{b^\kappa_\alpha}{b} - \delta^\kappa_\alpha\right) \ln \frac{2Z_\alpha + B^{\star\kappa}_\alpha (u_\alpha + \sqrt{u^2_\alpha - 4w_\alpha})}{2Z_\alpha + B^{\star\kappa}_\alpha (u_\alpha - \sqrt{u^2_\alpha - 4w_\alpha})}\end{aligned} \tag{14}$$

where $Z_\alpha = p_\alpha V_{m\alpha}/(RT_\alpha)$ is the *compressibility factor* of the phase, $A^\star_\alpha = a_\alpha p_\alpha/(RT_\alpha)^2$ and $B^\star_\alpha = b_\alpha p_\alpha/(RT_\alpha)$. Figure 2 visualizes the component fugacities of a binary water-methane mixture in the gas phase using the Peng-Robinson equation of state.

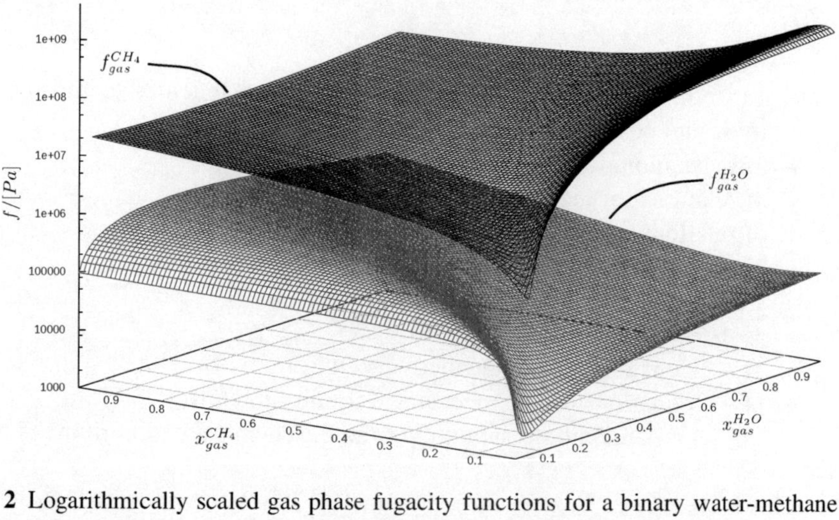

Fig. 2 Logarithmically scaled gas phase fugacity functions for a binary water-methane gas phase mixture at 275 *bar* pressure and temperature of 345 *K*. The Peng-Robinson equation of state was used in conjunction with equation (14) and equation (8) (with interaction coefficients 0). The parameters for the pure components are taken from [8].

3.5 Fluid-Mechanical Quantities

Finally, the purely fluid-mechanical quantities dynamic viscosity μ_α and relative permeability $k_{r\alpha}$ are assumed to be closed functions of the pressure, temperature and composition of all phases. For the dynamic viscosity, approaches like the one proposed by Chung et al., Reichenberg, Orrick and Erbar and Wilke are applicable [8]. The relative permeability $k_{r\alpha}$ is usually linked directly to the capillary pressure model.

3.6 A Book-Keeping Interlude

For the M-phase, N-component system, $M \cdot N + 7M + 3$ unknowns are required:

- 3 spatial parameters: gravity potential g, intrinsic permeability $\mathbf{K}$ and porosity ϕ
- M mean molar masses $\overline{M}_\alpha$
- M dynamic viscosities μ_α
- M relative permeabilities $k_{r\alpha}$
- M saturations S_α
- M temperatures T_α
- M pressures p_α
- M molar volumes $V_{m\alpha}$
- $N \cdot M$ unknowns for the phase compositions x_α^κ

Nota bene, that the temperatures T_α are not directly required by the differential equations for conservation of mass (3), but are necessary for thermodynamics.

On the other hand, $M \cdot N + 6M + 2$ orthogonal constraints have been identified so far:

- 3 externally specified spatial parameters (gravity potential g, intrinsic permeability $\mathbf{K}$ and porosity ϕ)
- M mean molar masses $\overline{M}_\alpha$
- M relations for the dynamic viscosities μ_α
- M relations for the relative permeabilities $k_{r\alpha}$
- N mass balance equations (3)
- 1 closure condition for the saturations (4)
- M thermal equations of state (6)
- $M - 1$ constraints from the thermal equilibrium (9)
- $M - 1$ equations for the mechanical equilibrium (10)
- $(M - 1) \cdot N$ conditions stemming from the chemical equilibrium (11)

This leaves $M + 1$ degrees of freedom which have to be specified via auxiliary constraints of the flow model. The next subsection will outline three approaches.

3.7 *Model Constraints*

First, temperature T is obtained as an external parameter if the model does not balance energy; If it does, in addition to the mass balance equations (3), it will include an additional partial differential equation for this purpose [5].

In both cases, M auxiliary constraints must still be provided.

Immiscibility

The first possibility which is discussed here is to assume immiscibility. For this, each phase is assumed to be composed of exactly one component, i.e. $M = N$ holds. For this, the fugacity coefficient is assumed to be finite for the component assigned to a phase and infinite for all other components, for example:

$$\Phi_\alpha^\kappa = \begin{cases} 1 & \text{if } \kappa = \alpha \\ \to \infty & \text{else} \end{cases} \tag{15}$$

As a consequence, the M auxiliary assumptions of models assuming immiscibility are

$$\sum_\kappa x_\alpha^\kappa = 1\,. \tag{16}$$

Locally Known Phase State

A second possibility is to assume that the local phase state is known at every given spatial location, i.e. that the fluid phases which are present are known. For each phase α which is present, the constraint

$$\sum_\kappa x_\alpha^\kappa = 1 \tag{17}$$

holds. If phase α is not present, it is assumed that the saturation of this phase is zero, i.e.

$$S_\alpha = 0\,. \tag{18}$$

This approach is commonly called *primary variable switching* [3] because the degrees of freedom are adapted to the local model assumptions which depend on the local thermodynamic state.

Complementarity Conditions

The final set of model assumptions outlined here requires that the fugacity of each component in every phase is strictly monotonous in regard to the component's mole fraction.

First, it is observed that under this precondition, a phase can only be present if the sum of the mole fractions of its constituents is 1, i.e.

$$S_\alpha > 0 \implies \sum_\kappa x_\alpha^\kappa = 1 \tag{19}$$

holds for each phase α. Also, if the sum of the mole fractions of a phase is smaller than 1, the phase cannot be present, which yields

$$\sum_\kappa x_\alpha^\kappa < 1 \implies S_\alpha = 0\,. \tag{20}$$

Together with the facts that a saturation cannot be negative and that each phase must be either present or not present, (19) and (20) imply

$$S_\alpha\left(1 - \sum_\kappa x_\alpha^\kappa\right) = 0, \quad 1 - \sum_\kappa x_\alpha^\kappa \geq 0, \quad S_\alpha \geq 0\,. \tag{21}$$

Conditions (21) can be captured using so-called non-linear complementarity functions, which can be used as for the missing M additional equations. For details, it is referred to [6].

4 Implementation in DuMu$^\mathrm{x}$

The following section is dedicated to an overview of how the abstract thermodynamic framework discussed in the previous section is implemented within the DuMu$^\mathrm{x}$ flow simulator.

4.1 Discretization of the Partial Differential Equations

The balance equations (3) are discretized using the box method described in [5], which is a vertex centered finite volume method. For the spatial discretization DuMu$^\mathrm{x}$ uses the DUNE-GRID interface extensively.

For temporal discretization, the implicit Euler method is used. The resulting implicit discrete non-linear system of equations is then solved using a damped Newton-Raphson iteration [9]. The linear system of equations obtained for each iteration of the Newton-Raphson method is solved using one of the preconditioned iterative block solvers provided by DUNE-ISTL. A thorough overview on DuMu$^\mathrm{x}$ can be found in [4].

As model constraints mentioned in section 3.7, immiscibility, primary variable switching, and the approach presented in [6] are implemented.

4.2 *The Material Law Framework*

This section is dedicated to the implementation of the thermodynamic constraints, which stem from the assumption of local thermodynamic equilibrium as outlined in section 3.4. It should be noted that, to some extend, this part of DuMu^x is still a work in progress and programming interfaces may change quickly.

Basic Assumptions

For models assuming thermodynamic equilibrium, all current DuMu^x models assume that the three sub equilibria may be calculated sequentially:

1. First the **thermal** equilibrium needs to be calculated. This step is usually quite simple, since the temperature T is either provided as a given parameter (for isothermal models), or it is a primary variable (for non-isothermal models).
2. Second, is supposed that the **mechanical** equilibrium may be treated. Since the chemical equilibrium is not yet calculated at this point, the capillary pressure may only depend on temperature and saturations, but not on phase composition. Calculating the capillary pressures may be quite challenging, depending on the choice of primary variables.
3. Finally, the algorithm for determining the **chemical** equilibrium may depend on the results of both, the thermal and the mechanical equilibrium calculations. In general, determination of the chemical equilibrium is by far the most complex part – in a multi-phase context, it usually requires inverting the component fugacity functions (13), which typically exhibit strongly non-linear fugacity coefficients, as illustrated by (14) for cubic equations of state.

The FluidState Concept

The central concept of the DuMu^x material law framework is the FluidState. A fluid state object provides access to all thermodynamic quantities required by the flow model. A fluid state is any class which, amongst others, provides the following methods. (Scalar is the C++ used floating point type, which is provided as a template argument):

- Scalar saturation(**int** phaseIdx) **const**
  The saturation of a fluid phase, [-]
- Scalar moleFrac(**int** phaseIdx, **int** compIdx) **const**
  The mole fraction of a component within a fluid phase [-]
- Scalar meanMolarMass(**int** phaseIdx) **const**
  The sum of all component mass fractions within a fluid phase $[kg/mol]$
- Scalar fugacity(**int** phaseIdx, **int** compIdx) **const**
  The fugacity of a component in a phase $[Pa]$
- Scalar molarVolume(**int** phaseIdx) **const**
  The molar volume of a fluid phase $[m^3/mol]$

- Scalar temperature(**int** phaseIdx) **const**
  The temperature of a fluid phase $[K]$
- Scalar pressure(**int** phaseIdx) **const**
  The pressure of a fluid phase $[Pa]$.

Note that this interface only allows to query quantities, but it does not make any assumptions of how they are stored internally. Currently, DuMu^x provides the GenericFluidState, and EquilibriumFluidState classes. The first stores all quantities which cannot trivially be converted amongst each other explicitly, the second assumes thermodynamic equilibrium and allows to save some space for storing the fugacities and temperatures.

Mechanical Equilibrium: Capillary Pressure Laws

Because of the basic assumption that the three constituent equilibria of thermodynamic equilibrium can be calculated in the order *thermal* $\rightarrow$ *mechanical* $\rightarrow$ *chemical*, capillary pressure laws currently can only be functions of temperature, phase saturations and phase pressures, but may not depend on phase compositions. The DuMu^x API for capillary pressure curves is based on the idea of separating the actual curves from their parameters. Thus, a capillary pressure law itself is a stateless class and only has the static members pC() which calculates all the capillary pressures and kr () calculating the relative permeabilities. Less abstractly, the two-phase law suggested by Brooks and Corey [1], is implemented like this:

```
template <class ScalarT,
          class wettingPhaseIdx,
          class nonwettingPhaseIdx,
          class ParamsT = BrooksCoreyParams<ScalarT> >
class BrooksCorey
{
public:
    typedef ParamsT Params;
    typedef typename Params::Scalar Scalar;

    template <class ScalarArray, class FluidState>
    static void pC(ScalarArray &dest,
                   const Params &params,
                   const FluidState &fs)
    {
      dest[wettingPhaseIdx] = 0;
      dest[nonwettingPhaseIdx] =
         params.pe(fs)*
         pow(fs.saturation(wettingPhaseIdx),
             - 1.0/params.lambda(fs));
    }

    template <class ScalarArray, class FluidState>
    static void kr(ScalarArray &dest,
                   const Params &params,
                   const FluidState &fs)
    {
      Scalar Sw = fs.saturation(wettingPhaseIdx);
      Scalar lambda = params.lambda(fs);

      dest[wettingPhaseIdx] = pow(Sw, 2.0/lambda + 3);

      Scalar exponent = 2.0/lambda + 1;
      Scalar tmp = 1. - Sw;
```

```
        dest[nonwettingPhaseIdx] =
            tmp*tmp*(1. - pow(Sw, exponent));
    }
};
```

With the parameter class defaulting to:

```
template <class ScalarT>
class BrooksCoreyParams
{
public:
    typedef ScalarT Scalar;

    // "entry pressure"
    template <class FluidState>
    Scalar pe(const FluidState &fs) const
    { return pe_; }

    // "shape" parameter
    template <class FluidState>
    Scalar lambda(const FluidState &fs) const
    { return lambda_; }

    void setPe(Scalar v)
    { pe_ = v; }
    void setLambda(Scalar v)
    { lambda_ = v; }
private:
    Scalar pe_;
    Scalar lambda_;
};
```

Fluid Systems

Fluid systems compute the quantities required to calculate the chemical equilibrium, as well as some purely fluid-mechanical quantities like dynamic viscosities and diffusion coefficients. Just like capillary pressure laws, fluid systems separate their parameters from the mathematical relations. In contrast to capillary pressure laws, fluid systems must provide two kinds of parameters:

- Static parameters, which provide all parameters that are independent of the thermodynamic state, i.e. they must be independent of the quantities provided by the fluid state. For example, these parameters include the critical pressure and temperature of the individual components, molar weights of the components or look-up tables.
- Also, fluid systems must provide mutable parameters, which are parameters that depend on the thermodynamic state. First of all, this means that the mutable parameters class must be a model of the `FluidState` interface, but it can also provide a cache for additional parameters which are specific for the chosen fluid system. For cubic equations of state, the additional mutable parameters include a_α^κ and b_α^κ as well as the a_α and b_α parameters required by the equations (7), (8) and (14).
- Finally, fluid systems provide a set of static methods which actually calculate the thermodynamic quantities $\langle \cdot \rangle_\alpha^\kappa$ depending on the mutable parameters.

In the DuMu^x code, the static parameters of a simple two-phase water-nitrogen fluid system are implemented like this:

```
template <class Scalar>
struct H2ON2StaticParameters {
    // phase indices. "numPhases" is mandatory, the
    // indices themselves are for convenience
    enum { numPhases = 2;
           lPhaseIdx = 0, // Index of the liquid phase
           gPhaseIdx = 1 }; // Index of the gas phase
    // component indices
    enum { numComponents = 2;
           H2OIdx = 0, // Index of water
           N2Idx = 1 }; // Index of nitrogen
    static const char* phaseName(int phaseIdx) {/*...*/};
    static const char* componentName(int componentIdx){/*...*/};
    static Scalar molarMass(int compIdx) {/*...*/};
    // Some fluid system specific parameters ...
};
```

If no additional parameters are required (or caching of intermediate results is not worthwhile), any class implementing the `FluidState` interface can be used for the mutable parameters. The actual fluid system is thus implemented as follows:

```
template <class Scalar>
struct H2ON2FluidSystem
  : public H2ON2StaticParameters<Scalar>
{
    typedef Dumux::H2ON2StaticParameters<Scalar>
            StaticParameters;
    typedef Dumux::GenericFluidState<Scalar,
                                     StaticParameters>
            MutableParameters;

    static Scalar
    computeMolarVolume(MutableParameters &params,
                       int phaseIdx) { /* ... */ };

    static Scalar
    computeFugacity(MutableParameters &params,
                    int phaseIdx,
                    int compIdx) { /* ... */ };

    static Scalar
    computeFugacityCoeff(MutableParameters &params,
                         int phaseIdx,
                         int compIdx) { /* ... */ };
    static Scalar
    computeViscosity(MutableParameters &params,
                     int phaseIdx) { /* ... */ };
    // ...
};
```

Constraint Solvers

Fluid states are not required to make sure that they are consistent with the thermodynamics specified by the fluid system. To achieve this consistency is the task of the thermodynamic *constraint solvers*. Constraint solvers assume that some quantities have been set and make the `FluidState` consistent with the thermodynamics of the fluid system. Currently, DuMu^x provides the following constraint solvers:

- `CompositionFromFugacities`: Given that the fugacities of all components in a phase $\{f_\alpha^\kappa\}$ are specified, this constraint solver determines the phase composition in terms of mole fractions $\{x_\alpha^\kappa\}$. In other words, this inverts the relations (13), (14). Since, in general, these relations are a non-linear system of N equations for each phase, a damped Newton-Raphson algorithm is used for this constraint solver.
- `ImmiscibleComponents`: This solver makes the fluid state consistent with the immiscibility assumption (15).
- `KnownPhaseState`: This solver makes the fluid state consistent with (17). In general this results in a non-linear system of size $M \cdot N$. Presently this solver is only implemented for ideal mixtures – i.e. that the fugacity coefficients $\{\Phi_\alpha^\kappa\}$ are independent of the phase compositions $\{x_\alpha^\kappa\}$.

Spatially Dependent Parameters

All parameters which are dependent on the spatial location – i.e. gravity g, porosity ϕ, intrinsic permeability $\mathbf{K}$, temperature T (for models which do not balance energy) as well as the parameter objects required by the capillary pressure law – are collected in a SpatialParameters object. The API of these object is specific to the chosen spatial discretization as well as – to a lesser extend – the chosen physical model.

5 Conclusion

To conclude this discourse, it can be said that physically accurate simulation of multi-phase multi-component flow in porous media is a challenging task. This essay tries to give a rough overview of the challenges involved with a special focus on the relevant parts of thermodynamics. Also, the fundamental ideas of how these concepts are implemented within the context of the DuMu^x toolbox are outlined. It should be mentioned that, due to the complexity of the underlying physics, the DuMu^x thermodynamic framework currently still is a work in progress to some extend, albeit it is already in a usable state.

References

1. Brooks, R.H., Corey., A.T.: Hydraulic properties of porous media. Tech. rep., Colorado State University, Fort Collins, Hydrology Paper No. 3, p. 27 (1964)
2. Class, H., Helmig, R.: Numerical simulation of non-isothermal multiphase multicomponent processes in porous media. Applications for the injection of steam and air. Adv. Water Resour. 25, 551–564 (2002)
3. Class, H., Helmig, R., Bastian, P.: Numerical simulation of non-isothermal multiphase multicomponent processes in porous media. An efficient solution technique. Adv. Water Resour. 25, 533–550 (2002)
4. Flemisch, B., Darcis, M., Erbertseder, K., Faigle, B., Lauser, A., Mosthaf, K., Müthing, S., Nuske, P., Tatomir, A., Wolff, M., Helmig, R.: Dumux: Dune for multi - {Phase, Component, Scale, Physics,...} flow and transport in porous media. SimTech-Preprint 2010-46, Universität Stuttgart (2010); submitted to Adv. Water Resour.

5. Helmig, R.: Multiphase Flow and Transport Processes in the Subsurface: A Contribution to the Modeling of Hydrosystems, 1st edn. Springer (1997)
6. Lauser, A., Hager, C., Helmig, R., Wohlmuth, B.I.: A new approach for phase transitions in miscible multi-phase flow in porous media. SimTech-Preprint 2010-34, Universität Stuttgart (2010); submitted to Adv. Water Resour.
7. Nolting, W.: Grundkurs theoretische Physik, 4th edn., vol. 6. Springer (2002)
8. Reid, R.C., Prausnitz, J.M., Poling, B.E.: The Properties of Gases and Liquids, 4th edn. McGraw-Hill (1987)
9. Sebah, P., Gourdon, X.: Newton's method and high order iterations. Tech. rep., Numbers, Constants and Computation (2001), `http://numbers.computation.free.fr/Constants/Algorithms/newton.ps`
10. van Genuchten, M.T.: A closed-form equation for predicting the hydraulic conductivity of unsaturated soils. Journal of the Soil Science Society of America 44(5), 892–898 (1980)

Using DUNE-PDELAB for Two-Phase Flow in Porous Media

Christoph Grüninger

Abstract. In this work, two-phase flow in porous media is simulated numerically with Discontinuous Galerkin methods. The Symmetrical Interior Penalty Galerkin method (SIPG), the Non-symmetrical Interior Penalty Galerkin method (NIPG) and the scheme from Oden, Babuška and Baumann (OBB) are considered. The terminology and the examples are taken from soil science. First, the Richards equation is solved, then a two-phase flow problem in the pressure-saturation formulation. The results of a final year project in computer science are presented and it is shown what a student can achieve with DUNE-PDELAB within six months.

1 Introduction

The objective of this work is to write fast simulation programs with today's mathematical methods and an existing software toolkit as a final year project (Diplomarbeit) in computer science. Within six months this should be done using DUNE-PDELAB. The interdisciplinary components from scientific computing, such as the conceptual or the mathematical model, are also discussed because they form the basic principles for the implemented numerical model.

Knowledge of water movement can both improve the exploitation of limited water reserves and protect as much water as possible for future usage. The mechanisms of groundwater movement must be known to properly specify water protection areas and anticipate potential dangers arising from pollution. In many cases two phases are considered, e. g. groundwater and soil air or petroleum and water. A recent example is given by Schwartz [14] where a German underground storage facility for nuclear waste is simulated with the two phases water and hydrogen.

Christoph Grüninger
Lehrstuhl für Hydromechanik und Hydrosystemmodellierung, Universität Stuttgart, Pfaffenwaldring 61, Stuttgart, 70569, Germany
e-mail: `christoph.grueninger@iws.uni-stuttgart.de`

In general, the mathematical models used in this work lead to problems which cannot be solved analytically. Numerical methods can find approximations but the methods have to face many difficulties like non-linear functions, discontinuities, singularities or large systems of linear equations. The implementation challenges are complicated meshes with the need for local refinement, heavy memory consumption and complicated software that has been growing over time.

Discontinuous Galerkin (DG) methods are successfully used in many fields like fluid dynamics [5] or free-surface water flow [8]. They provide local conservation of mass and energy, higher order convergence rates, flexibility in terms of meshes and are capable of *hp*-refinement; they are promising for subsurface flow, too. The DG methods used in this work are the Symmetrical Interior Penalty Galerkin method (SIPG), the Non-symmetrical Interior Penalty Galerkin method (NIPG) and the scheme from Oden, Babuška and Baumann (OBB). An overview of DG methods is given in [1].

Section 2 introduces the relevant physical effects and mathematical models. In Section 3 discretizations with the DG methods are described. First numerical results with the Richards equation are shown in Section 4. Then Section 5 presents numerical results obtained from a two-phase flow solver using the pressure-saturation formulation. Finally, the conclusions are drawn in Section 6.

2 Flow in Porous Media

This section introduces the terminology and the models used in simulations of flow and transport in porous media. First porous media are defined and relevant effects are described, following [2]. Then a mathematical model for two-phase flow is given.

2.1 Terms and Models

A porous medium is a heterogeneous body with small enclosed pores. They are connected and filled with one or more fluids which are able to move through the pores. In the following, only the case of two immiscible fluids will be considered, called phases. Such flow through porous media occurs in petroleum geology where water and oil flow in soil problems are studied aiming at oil exploitation. In soil science, groundwater and subsurface air are the fluids being historically studied, e. g. in [12].

A hypothetical continuum on the macroscopic scale is used instead of discrete molecules or a simulation of single pores. Mass, density ρ and pressure p are considered as continuous quantities, velocity u and porosity ϕ as discontinuous quantities. The porosity ϕ is the ratio of the pore space to the overall volume. The ratio of space occupied by one fluid phase to the pore space is called saturation S. In a two-phase system the pore space must be filled with either one of the phases α or β

leading to $S_\alpha + S_\beta = 1$. In most cases a portion of each fluid is trapped in small or dead end pores, so the α-phase residual saturation $S_{\alpha r}$ gives α's minimal saturation. The effective saturation $\bar{S}_\alpha = {(S_\alpha - S_{\alpha r})}/{(1-S_{\alpha r})}$ describes the saturation excluding the residual part of the phase α.

The gravitational acceleration g is a force pointing downwards. Without any water-movement, in the hydrostatic case, the linear law $p = -\rho g z$ is valid for the pressure p. The depth z is negative in the saturated zone, zero at the groundwater table and positive above. For groundwater flow the pressure differs from the hydrostatic case by the hydraulic pressure $p + \rho g z$. The piezometric head $h = p/\rho g + z$ is the same variable but interpreted as a depth, thus measured in meter.

For every pair of fluids in a porous medium, one is the wetting and the other is the non-wetting phase. For example, water is the wetting phase if the second one is oil. The fluid with the higher affinity to the solid is the wetting one. From now on, the fluids are called wetting and non-wetting phases with subscripts w and n to label corresponding variables. If something applies for both phases the subscript $\alpha \in \{w, n\}$ is used. On the boundary surface between two fluids, a pressure difference occurs which is called capillary pressure p_c. The capillary pressure on the pore level is replaced by a macroscopic pressure difference $p_c = p_n - p_w$ which is always non-negative by definition. It can be approximated with functions depending on the effective saturation $\bar{S}_w$. The capillary pressure may become very large and steep for small effective saturations $\bar{S}_w$ which leads to reduced numerical accuracy and higher solver costs.

The mobility $\lambda_\alpha(S_\alpha) = k_{r\alpha}(S_\alpha)/\mu_\alpha$ is a combination of properties from the fluid and the solid matrix. The dynamic viscosity μ_α describes friction inside the fluid. The relative permeability $k_{r\alpha}(S_\alpha)$ describes the interference of flow by the other phase and depends on the effective saturations $\bar{S}_\alpha$. The intrinsic permeability K is a macroscopic simplification of the soil's microscopic grain structure. It is modeled as a location dependent positive symmetric tensor.

The velocities of the fluids are treated as quantities on a suitable volume V. Darcy's law states a proportional relationship between the velocity and the pressure gradient

$$u_\alpha = -\frac{k_{r\alpha}(S_\alpha)}{\mu_\alpha} K (\nabla p_\alpha - \rho_\alpha g) = -\lambda_\alpha(S_\alpha) K (\nabla p_\alpha - \rho_\alpha g). \qquad (1)$$

It is a special case of the Navier-Stokes equation and can be regarded as a momentum conservation law.

The conservation of mass for each phase α,

$$\frac{\partial}{\partial t}(\phi S_\alpha \rho_\alpha) + \nabla \cdot (\rho_\alpha u_\alpha) = q_\alpha, \qquad (2)$$

can be obtained from a specialization of the Reynolds transport theorem, cf. [11].

2.2 Darcy Two-Phase Flow

With the substitution of u in (2) by Darcy's law (1) the problem description of the two-phase flow,

$$\frac{\partial}{\partial t}(\phi \rho_\alpha S_\alpha) - \nabla \cdot \left(\rho_\alpha \lambda_\alpha (S_\alpha) K (\nabla p_\alpha - \rho_\alpha g) \right) = q_\alpha, \qquad \alpha \in \{w, n\}, \tag{3}$$

has four unknowns: two saturations S_α and two pressures p_α. Using the sum of saturations $S_w + S_n = 1$ we can express S_w by S_n. Similarly we eliminate p_n with the capillary pressure $p_c(S_w) = p_n - p_w$. The formulation has the unknowns p_w and S_n left and is called pressure-saturation formulation:

$$\frac{\partial}{\partial t}(\phi \rho_w (1 - S_n)) - \nabla \cdot \left(\rho_w \lambda_w (1 - S_n) K (\nabla p_w - \rho_w g) \right) = q_w \tag{4}$$

$$\frac{\partial}{\partial t}(\phi \rho_n S_n) - \nabla \cdot \left(\rho_n \lambda_n (S_n) K (\nabla p_w + \nabla p_c (1 - S_n) - \rho_n g) \right) = q_n. \tag{5}$$

This is a system of hyperbolic-parabolic equations which will be solved fully implicitly in the following. Dirichlet and Neumann boundary conditions are described by functions g and j.

The four equations can be combined in various other ways, too. An overview of these formulations is given in [2].

2.3 The Richards Equation

With the phases water and air in the unsaturated zone, one can assume a constant air pressure. When the soil air is connected to the air above the ground, and water percolates slowly enough for air bubbles to escape upwards, soil air has atmospheric pressure. By scaling the atmospheric pressure to $p_n = 0$, the water pressure becomes $p_w = -p_c$. With the incompressibility of water, density ρ_w is constant. Then exists a saturation function $S(p_w) := p_c^{-1}(-p_w)$ and we can divide pressure p_w by $\rho_w g$ and use piezometric head h instead. The problem simplifies to the equation

$$\frac{\partial}{\partial t} S(h) - \nabla \cdot \left(K(h) (\nabla h - z) \right) = q \tag{6}$$

with $K(h)$ the absolute permeability. The equation is a non-linear form of the parabolic heat equation and was first published in [12].

3 Discontinuous Galerkin Scheme

The domain Ω is partitioned to $E = \{e_1, \ldots, e_{n_{\Delta x}}\}$ by a grid of sufficiently regular polygons in the usual way, see for example [4]. Between every pair of neighboring elements e_i and e_j is a face $f = e_i \cap e_j$. The set of faces is F and $|f|$ the length of f. The sets F_D and F_N contain element faces on the domain boundary with Dirichlet or Neumann conditions. Every face f has an arbitrary inside element e^- and outside element e^+. The face normal ν points from the inside to the outside element, in case of boundary faces ν is the outer normal. The evaluation of a function φ on a face f depending on the side is $\varphi^{\pm}(x) := \varphi(x)|_{e^{\pm}}$. We define the jump $[\varphi](x) := \varphi^-(x) - \varphi^+(x)$ and the average $\langle\varphi\rangle(x) := 1/2\,(\varphi^-(x) + \varphi^+(x))$ of φ on a face.

The discontinuous finite element space is $V^k(E) = \{\psi \in L^2(\Omega) \,|\, \forall e \in E : \psi|_e \in P_k(e)\}$ where k is a given integer and P_k is the space of polynomial functions of maximal total degree k. A function $\varphi \in V^k(E)$ may be discontinuous across faces.

To obtain the weak formulation of (2) we multiply every term with a test function v_α, integrate by parts per element and sum over all elements:

$$\sum_{e\in E}\int_e \frac{\partial}{\partial t}(\phi S_\alpha \rho_\alpha)\, v_\alpha \,\mathrm{d}x - \sum_{e\in E}\int_e (\rho_\alpha u_\alpha)\cdot \nabla v_\alpha \,\mathrm{d}x + \sum_{f\in F}\int_f [\rho_\alpha u_\alpha \cdot \nu\, v_\alpha]\,\mathrm{d}s$$
$$= \sum_{e\in E}\int_e q_\alpha v_\alpha \,\mathrm{d}x \qquad \forall v_\alpha \in V^k$$

We can rewrite the jump over faces because of a sufficiently regular exact solution,

$$\sum_{f\in F}\int_f [\rho_\alpha u_\alpha \cdot \nu\, v_\alpha]\,\mathrm{d}s = \sum_{f\in F}\int_f \langle \rho_\alpha u_\alpha \cdot \nu\rangle [v_\alpha]\,\mathrm{d}s. \tag{7}$$

To improve the stability, the term $\int_f [u_\alpha]\langle \rho_\alpha v_\alpha \cdot \nu\rangle\,\mathrm{d}s$ is added for every face f. This is the scheme from Oden, Babuška and Baumann (OBB) which is stable for $k \geq 2$. Now we substitute u_α with Darcy's law (1). We approximate both unknowns p_w and S_n with the discontinuous functions of degree k.

The time is discretized in steps of length Δt. We define any variable with a superscript $\cdot^i$ as the variable at the time $i\Delta t$. Using OBB we obtain a system of ordinary differential equations which is numerically solved with an implicit Euler scheme and Newton's method.

Altogether the problem is to find p_w^{i+1} and S_n^{i+1} with given p_w^i and S_n^i such that for all $(v_p \times v_S) \in V^k \times V^k$ are satisfied:

Pressure equation / wetting phase:

$$\frac{1}{\Delta t}\int_{e\in E} \phi\,\left(\rho_w^i(1-S_n^i) - \rho_w^{i-1}(1-S_n^{i-1})\right) v_p \,\mathrm{d}x + \int_{e\in E} \rho_w^i \lambda_w^i K(\nabla p_w^i - \rho_w^i g)\cdot \nabla v_p \,\mathrm{d}x$$
$$+ \int_{f\in F} \varepsilon \langle \rho_w^i \lambda_w^i K \nabla v_p \cdot \nu\rangle [p_w^i] - \langle \rho_w^i \lambda_w^i K \nabla p_w^i \cdot \nu\rangle [v_p] + \frac{\sigma}{|f|^\beta}[p_w^i][v_p]\,\mathrm{d}s$$
$$+ \int_{f\in F_{D,w}} \varepsilon \rho_w^i \lambda_w^i K \nabla v_p \cdot \nu (p_w^i - g_{p_w}^i) - \rho_w^i \lambda_w^i K \nabla p_w^i \cdot \nu v_p + \frac{\sigma}{|f|^\beta}(p_w^i - g_{p_w}^i) v_p \,\mathrm{d}s$$
$$- \int_{e\in E} q_w^i v_p \,\mathrm{d}x + \int_{f\in F_{N,w}} j_w^i v_p \,\mathrm{d}s = 0$$

On the Implementation of a Heterogeneous Multi-scale Finite Element Method for Nonlinear Elliptic Problems

Patrick Henning and Mario Ohlberger

Abstract. In this contribution, we formulate a heterogeneous multiscale finite element method (HMM) for monotone elliptic operators. This is done in the general concept of HMM, which was initially introduced by E and Engquist [E, Engquist, *The heterogeneous multiscale methods*, Commun. Math. Sci., 1(1):87–132, 2003]. Since the straightforward formulation is not suitable for a direct implementation in the nonlinear setting, we present a corresponding algorithm, which involves the computation of additional cell problems. The algorithm is validated by numerical experiments and can be used to effectively determine homogenized solutions.

1 Introduction

In this work, we are concerned with monotone elliptic multiscale problems of the following structure:

$$\begin{aligned} -\nabla \cdot A^{\varepsilon}(x, \nabla u^{\varepsilon}) &= f \ \text{ in } \ \Omega, \\ u^{\varepsilon} &= 0 \ \text{ on } \ \partial\Omega. \end{aligned} \tag{1}$$

Here, A^{ε} describes a rapidly oscillating, nonlinear diffusion operator, where the parameter ε represents the size of the microscale. For instance, if the oscillations are periodic, then ε denotes the period. These types of equations describe a large variety of engineering problems, such as transport of pollutants in groundwater or the conductivity of compositional materials. Resolving the entire microstructure, in order to perform a highly expensive fine-scale computation, typically results in an intractable

Patrick Henning · Mario Ohlberger
Institut für Numerische und Angewandte Mathematik, Einsteinstraße 62, 48149 Münster, Germany
e-mail: patrick.henning@math.uni-muenster.de, mario.ohlberger@uni-muenster.de

computational demand. To reduce the complexity of the problem significantly, equations such as (1) are often treated by so called multiscale methods, which include a large range of different approaches. In this work, we restrict ourselves to a realization of the *heterogeneous multiscale finite element method* (HMM). The concept of HMM was initially introduced by E and Engquist in 2003 [5] and extended by several contributions. A good overview on this can be found in the work of Abdulle [1]. A HMM realization with near optimal computational complexity is given by Abdulle and Engquist [2], a version for the treatment of advection-diffusion problems with large drift by Henning and Ohlberger [9]. A-posteriori error estimates are derived in the work of Henning and Ohlberger [11, 7, 8]. In the framework of Multiscale finite element methods (MsFEM) Efendiev, Hou and Ginting [6] introduce a multiscale finite element method for nonlinear problems. The HMM for monotone operators, that we present in this contribution, is formulated in analogy to the linear case as presented in [5]. However, if we are interested in implementing the method, we face the problem, that the formulation is not yet suitable for programming purposes since it is not clear how to efficiently compute the Jacobian matrix in a Newton scheme. In the last part of Section 3 we discuss this issue in detail. In fact, we figure out that the usage of an accessory Newton method results in the assembling of additional cell problems. Considering these new problems, we present a general algorithm for solving the discrete HMM problem for monotone operators. This algorithm is implemented using the software toolbox DUNE, which is for instance introduced in [3]. In particular, we employ the DUNE module DUNE-FEM (see [4]).

2 Analytical and Discrete Setting

Let $\Omega \subset \mathbb{R}^N$ denote a polygonal bounded domain with dimension $N \leq 3$. By $\mathring{H}^1(\Omega)$ we define the space of $H^1(\Omega)$-functions with compact support in Ω. In the following, we are concerned with solving the following elliptic multiscale problem:

Problem 1 (Nonlinear elliptic equation with fast oscillations). Find $u^\varepsilon \in \mathring{H}^1(\Omega)$:

$$\int_\Omega A^\varepsilon(x, \nabla u^\varepsilon) \cdot \nabla \Phi(x)\, dx = \int_\Omega f(x)\Phi(x)\, dx \qquad \forall \Phi \in \mathring{H}^1(\Omega). \tag{2}$$

Here, $A^\varepsilon : \Omega \times (L^2(\Omega))^N \to (L^2(\Omega))^N$ denotes a nonlinear diffusion operator, which meets the subsequent monotonicity and continuity conditions: there exist two constants $0 < \alpha \leq \beta < \infty$ such that for all $x \in \Omega$ and all $\xi_1, \xi_2 \in \mathbb{R}^N$:

$$\begin{aligned} < A^\varepsilon(x,\xi_1) - A^\varepsilon(x,\xi_2), \xi_1 - \xi_2 > &\geq \alpha|\xi_1 - \xi_2|^2, \\ |A^\varepsilon(x,\xi_1) - A^\varepsilon(x,\xi_2)| &\leq \beta|\xi_1 - \xi_2| \text{ and} \\ A^\varepsilon(x,0) &= 0. \end{aligned}$$

With these assumptions, we have a unique solution of problem (2) (see e.g. [10]).

In order to create a suitable discrete setting, in which we can formulate the heterogeneous multiscale method, we denote by $\mathscr{T}_H$ a regular simplicial partition of Ω with elements T. x_T is the barycenter of $T \in \mathscr{T}_H$. By $Y = (-\frac{1}{2}, \frac{1}{2})^N$ we define the 0-centered unit cube. Scaling Y with $\kappa \in \mathbb{R}_{>0}$ and shifting it by $x_T \in T$, we introduce the notation $Y_{T,\kappa} := \{x + \kappa y | \; y \in Y\}$. By $\mathscr{T}_h$, we denote a regular, periodic partition of Y. y_K is the barycenter of $K \in \mathscr{T}_h$. The mapping $x_T^\kappa : Y \to Y_{T,\kappa}$ is given by $x_T^\kappa(y) := x_T + \kappa y$. Moreover, we introduce the following discrete spaces, where $\mathbb{P}^1$ is the space of polynomials of degree 1:

$$V_H(\Omega) := \{\Phi_H \in \mathring{H}^1(\Omega) \cap C^0(\Omega) \; |\Phi_{H_{|T}} \in \mathbb{P}^1(T) \; \forall T \in \mathscr{T}_H\};$$

$$W_h(Y) := \{\phi_h \in C^0(Y) | \; \phi_h \text{ is } Y\text{-periodic, } \int_Y \phi_h = 0 \text{ and } \phi_{h|K} \in \mathbb{P}^1(K) \, \forall K \in \mathscr{T}_h\};$$

$$W_h(Y_{T,\kappa}) := \{\phi_h \in C^0(Y_{T,\kappa}) | \; (\phi_h \circ x_T^\kappa) \in W_h(Y)\}.$$

Definition 1 (Reconstruction operator). In order to locally reconstruct information of the fine-scale behavior of u^ε, we introduce the *local reconstruction operator* $R_h^T : V_H(\Omega) \to V_H(\Omega) + W_h(Y_{T,\delta})$. For $\delta > 0$ and $\Phi_H \in V_H(\Omega)$, the corresponding reconstruction $R_h^T(\Phi_H) \in \Phi_H + W_h(Y_{T,\delta})$ is defined as the solution of

$$\int_{Y_{T,\delta}} A^\varepsilon \left(x, \nabla_x R_h^T(\Phi_H)(x)\right) \cdot \nabla_x \phi_h(x) \; dx = 0 \qquad \forall \phi_h \in W_h(Y_{T,\delta}). \tag{3}$$

Note that R_h^T is well defined since (3) always has a unique solution, if $Y_{T,\delta} \subset T$. We also point out that (3) is the discrete analog to the well-posed so-called cell problems known from standard homogenization theory (cf. [12]).

3 The Heterogeneous Multiscale Finite Element Method for Monotone Operators

The general idea behind the concept of HMM is based on a scale separation of the solution of problem (2). If we assume, that u^ε can be separated into its coarse- and fine-scale part, we can artificially separate the test functions in the same way and postulate $\Phi = \Phi^{coarse} + \phi^{fine}$, where ϕ^{fine} remains very small in the L^2-norm, but produces large gradients due to fast oscillations. Together with (2) and choosing $\phi^{fine} = 0$ we obtain

$$\int_\Omega A^\varepsilon(\cdot, \nabla u^\varepsilon) \cdot \nabla \Phi^{coarse} = \int_\Omega f \Phi^{coarse}. \tag{4}$$

On the other hand, with $\Phi^{coarse} = 0$ we get:

$$\int_\Omega A^\varepsilon(\cdot, \nabla u^\varepsilon) \cdot \nabla \phi^{fine} = \int_\Omega f \phi^{fine} \approx 0. \tag{5}$$

Discretizing (4) and (5), restricting the fine-scale computations on representative cells $x_T + \delta Y$ and reconstructing the macroscopic test function in order to guarantee uniqueness and existence, gives us the canonical formulation of HMM, as initially suggested by E and Engquist [5] for linear elliptic equations. In analogy, we can state the heterogeneous multiscale method for monotone operators:

Definition 2 (Heterogeneous Multiscale Method for monotone operators). We define the HMM approximation u_H of u^ε by: $u_H \in V_H(\Omega)$ solves

$$(f_H, \Phi_H)_{L^2(\Omega)} = A_H(u_H, \Phi_H) \qquad \forall \Phi_H \in V_H(\Omega), \tag{6}$$

with

$$A_H(u_H, \Phi_H) := \sum_{T \in \mathcal{T}_H} |T| \fint_{Y_{T,\varepsilon}} A^\varepsilon\left(x, \nabla_x R_h^T(u_H)(x)\right) \cdot \nabla_x R_h^T(\Phi_H)(x)\ dx. \tag{7}$$

Here, R_h^T denotes the local reconstruction operator, as it has been is defined in (3). For the oversampling parameter δ, we assume $\delta \geq \varepsilon$. An expedient choice for the periodic case could be $\delta = \varepsilon$, for the non-periodic case $\delta = 2\varepsilon$.

Note, for every $\delta \geq \varepsilon$, the HMM produces a unique solution $u_H \in V_H(\Omega)$. This is a direct effect of the monotonicity of $A^\varepsilon(x,\cdot)$. For the case that $A^\varepsilon(x,\cdot)$ is a linear operator, (7) results in a linear system of equations, which is symmetric and therefore cheap to solve.

If we assume, that the nonlinear diffusion operator A^ε only contains periodic oscillations, the homogenization of problem (2) is well known. However, even in this scenario, it is not immediately clear, how to carry over the analytical results to a suitable discretization. That is why the heterogeneous multiscale finite element method above becomes an effective scheme to determine the homogenized solution of elliptic multiscale problems with a periodically oscillating, monotone diffusion operator. In fact, with slight modifications, we can show, that the HMM is equivalent to a straightforward discretization of the standard homogenized equation of (2).

Problems Concerning the Implementation

In order to determine the effect of the reconstruction operator R_h^T on a function Φ_H, we need to solve the problems given by (3). Each of these so-called cell problems is a standard nonlinear elliptic problem, which can be solved easily by common methods. However, the situation is completely different for the HMM macro problem. In order to solve (6) numerically, we might face difficulties, depending on the structure of the nonlinearity. This is due to the fact that we do not only have one nonlinearity produced by A^ε, but an additional nonlinearity within, produced by the reconstruction operator R_h^T. In contrast to A^ε, an evaluation of R_h^T is not cheap, which is why we require a special treatment for this equation. This claim will be emphasized, if we have a closer look at (6). Let $\Phi_1, ..., \Phi_M$ denote the Lagrange base of V_H. Defining $G : \mathbb{R}^M \to \mathbb{R}^M$ by

$$(G(\alpha))_i := \sum_{T\in\mathscr{T}_H} |T| \fint_{Y_{T,\varepsilon}} A^\varepsilon(x, \nabla_x R_h^T(\sum_j^M \alpha_j\Phi_j)(x)) \cdot \nabla_x R_h^T(\Phi_i)(x)\ dx$$
$$- \int_\Omega f(x)\Phi_i(x)\,dx,$$

we are looking for $\bar{\alpha}$ with $G(\bar{\alpha}) = 0$. Solving this numerically (with an iterative scheme) requires either the computation of the Jacobian matrix $D_\alpha G$ or a corresponding approximation. There are two possibilities:

1. If we want to make direct use of $D_\alpha G$, we require the effect of the Fréchet derivative of the reconstruction operator, i.e. DR_h^T, which we can not determine explicitly.
2. If we approximate $D_\alpha G$ by difference quotients $\frac{(G(\alpha+he_j)-G(\alpha))_i}{h}$, we do not only need to determine $R_h^T(\sum_j^M \alpha_j\Phi_j)$, but also $R_h^T(\sum_j^M(\alpha_j+\sigma_{ji}h)\Phi_j)$ for any relevant i. Each of this involves solving a nonlinear elliptic problem for every $T \in \mathscr{T}_H$. Moreover this has to be done for every iteration step carried out for finding the solution of $G(\bar{\alpha}) = 0$. Therefore, a very high computational demand is implied.

In any case, we are dealing with accessory problems. In the next section we present a corresponding solution.

4 Implementation

In this section we want to state an algorithm in order to compute the solution of the HMM macro problem given by equation (6). This algorithm is a combination of HMM and Newton Method, which in particular can be used for solving the nonlinear elliptic multiscale problem (2). It is derived and stated in Subsection 4.1. In Subsection 4.2 we comment on its implementation in the DUNE framework.

4.1 HM Newton Method

We now consider a general treatment of the HMM macro problem, given by equation (6). First of all we emphasize that this treatment strongly depends on the structure of A^ε and should be modified according to problem specific characteristics. For instance, if we are dealing with a linear diffusion operator, i.e. $A^\varepsilon(x,\xi) = \bar{A}^\varepsilon(x)\xi$ with a matrix $\bar{A}^\varepsilon$, the implementation of (6) is straightforward, since (6) directly yields a linear system of equations. For the case that the nonlinearity is of the form $A^\varepsilon(x,\xi) = \bar{A}^\varepsilon(x,\xi)\xi$, with uniformly coercive $\bar{A}^\varepsilon$, the exact problem (2) can be linearized by means of the iteration $\int_\Omega \bar{A}^\varepsilon(x,\nabla u^{n-1})\nabla u^n \nabla\Phi = \int_\Omega f\Phi$, where u^n converges to u^ε under appropriate assumptions. Again, solving the HMM macro problem is uncomplicated, since we can use the linearized form above.

In the situation of a very general nonlinearity, we are facing problems which we described at the end of Section 3. Therefore, it is necessary to derive a method, which explicitly incorporates a general way for solving the macro problem (6). In the following we present an approach, where we combine the HMM with a Newton method. This procedure results in additional cell problems to solve.

First of all, we want to introduce a new representation of equation (6), which is more convenient with regard to applying the Newton method. For this purpose, we define the local correction operator Q_h^T.

Definition 3 (Correction operator). With the local reconstruction operator R_h^T defined in (3), we introduce a corresponding correction operator Q_h^T, which only yields (scaled) microscopic contributions. The operator $Q_h^T : V_H(\Omega) \to W_h(Y)$ is defined by $Q_h^T(\Phi_H)(y) := \frac{1}{\delta}\left(R_h^T(\Phi_H) - \Phi_H\right) \circ x_T^\delta(y)$ for $\Phi_H \in V_H(\Omega)$.

With this definition, we obtain the subsequent representation of equation (6):

Lemma 1 (Reformulation of the HMM operators). *For $T \in \mathcal{T}_H$, $\Phi_H \in V_H(\Omega)$ and Q_h^T given by Definition 3, we have that $Q_h^T(\Phi_H)$ is the solution of the following cell problem:*

$$\int_Y A^\varepsilon\left(x_T + \delta y, \nabla_x \Phi_H(x_T) + \nabla_y Q_h^T(\Phi_H)(y)\right) \cdot \nabla_y \phi_h(y)\, dy = 0 \quad \forall \phi_h \in W_h(Y). \tag{8}$$

Moreover, for $\Psi_H, \Phi_H \in V_H(\Omega)$, we can rewrite A_H (given by (7)) as:

$$\begin{aligned} A_H(\Psi_H, \Phi_H) = \sum_{T \in \mathcal{T}_H} |T| \fint_{Y_{\frac{\varepsilon}{\delta}}} & A^\varepsilon(x_T + \delta y, \Psi_H(x_T) + \nabla_y Q_h^T(\Psi_H)(y)) \\ & \cdot \left(\nabla_x \Phi_H(x_T) + \nabla_y Q_h^T(\Phi_H)(y)\right)\; dy. \end{aligned} \tag{9}$$

Proof. The results are directly obtained by using the transformation formula and the following equation, which holds for every $\Phi_H \in V_H$:

$$\begin{aligned} \nabla_y Q_h^T(\Phi_H)(y) &= \frac{1}{\delta} \nabla_y \left(R_h^T(\Phi_H)(x_T + \delta y) - \Phi_H(x_T + \delta y)\right) \\ &= \left(\nabla R_h^T(\Phi_H)\right)(x_T + \delta y) - \nabla \Phi_H(x_T). \end{aligned}$$

□

Now, we are prepared to derive the HM Newton algorithm. By $\{\Phi_i |\; 1 \le i \le M\}$ we denote the Lagrange base of the discrete space $V_H(\Omega)$. In particular M defines the number of macroscopic base functions. If we want to solve the HMM macro problem, i.e. find $u_H \in V_H$ with

$$A_H(u_H, \Phi_i) - (f, \Phi_i)_{L^2(\Omega)} = 0 \qquad \forall \Phi_i,\; 1 \le i \le M,$$

we can use the reformulation (9), so that we can equivalently look for $\bar{\alpha} \in \mathbb{R}^M$ with $G(\bar{\alpha}) = 0$. Here, $G : \mathbb{R}^M \to \mathbb{R}^M$ is defined by

$$(G(\alpha))_i := \sum_{T\in\mathscr{T}_H} |T| \fint_{Y_{\frac{\varepsilon}{\delta}}} A^\varepsilon(x_T+\delta y, \sum_{j=1}^M \alpha_j \nabla_x \Phi_j(x_T) + \nabla_y Q_h^T(\sum_{j=1}^M \alpha_j \Phi_j)(y))$$
$$\cdot \left(\nabla_x \Phi_i(x_T) + \nabla_y Q_h^T(\Phi_i)(y)\right)\ dy - \int_\Omega f(x)\Phi_i(x)\,dx.$$

If $\bar{\alpha}$ is computed, so is u_H, where we have the relation $u_H(x) = \sum_{j=1}^N \bar{\alpha}_j \Phi_j(x)$.

In order to solve the nonlinear algebraic equation $G(\bar{\alpha}) = 0$, we want to use the Newton method, with which we get the following iteration scheme:

$$\alpha^{(n+1)} = \alpha^{(n)} - \left((D_\alpha G)(\alpha^{(n)})\right)^{-1} G(\alpha^{(n)}).$$

Here, $D_\alpha G$ is the Jacobian matrix of G. With $\triangle\alpha^{(n)} := \alpha^{(n+1)} - \alpha^{(n)}$ we obtain:

$$D_\alpha G(\alpha^{(n)})\triangle\alpha^{(n)} = -G(\alpha^{(n)}).$$

This implies, that we need to compute the components of $D_\alpha G$. To determine $\frac{\mathrm{d}}{\mathrm{d}\alpha_k}(G(\alpha))_i$ for any $\alpha \in \mathbb{R}^M$, we calculate:

$$D_{\alpha_k} A^\varepsilon(x_T+\delta y, \sum_{j=1}^M \alpha_j \nabla_x \Phi_j(x_T) + \nabla_y Q_h^T(\sum_{j=1}^M \alpha_j \Phi_j)(y))$$
$$= D_\xi A^\varepsilon(x_T+\delta y, \sum_{j=1}^M \alpha_j \nabla_x \Phi_j(x_T) + \nabla_y Q_h^T(\sum_{j=1}^M \alpha_j \Phi_j)(y))$$
$$(\nabla_x \Phi_k(x_T) + \nabla_y (D_{\alpha_k}(Q_h^T(\sum_{j=1}^M \alpha_j \Phi_j)))(y))$$

It remains to determine $(D_{\alpha_k}(Q_h^T(\sum_{j=1}^M \alpha_j \Phi_j)))(y)$, which depends on the effect of the Fréchet derivative of the correction operator. Here, we need to use the reformulated cell problem (8) to obtain

$$0 = D_{\alpha_k} \int_Y A^\varepsilon(x_T+\delta y, \sum_{j=1}^M \alpha_j \nabla_x \Phi_j(x_T) + \nabla_y Q_h^T(\sum_{j=1}^M \alpha_j \Phi_j)(y)) \cdot \nabla_y \phi_h(y)\,dy$$
$$= \int_Y (D_\xi A^\varepsilon)(x_T+\delta y, \sum_{j=1}^M \alpha_j \nabla \Phi_j(x_T) + \nabla_y Q_h^T(\sum_{j=1}^M \alpha_j \Phi_j)(y))$$
$$(\nabla_x \Phi_k(x_T) + \nabla_y (D_{\alpha_k}(Q_h^T(\sum_{j=1}^M \alpha_j \Phi_j)))(y)) \cdot \nabla_y \phi_h(y)\,dy.$$

Since this equation holds for every $\phi_h \in W_h(Y)$, we have a characterization of $D_{\alpha_k}(Q_h^T(\sum_{j=1}^M \alpha_j \Phi_j))$. Combining the results, we can compute $D_\alpha G(\alpha^{(n)})$ by first solving (8) to obtain $Q_h^T(u_H^{(n)})$ and using this to determine $D_{\alpha_k}(Q_h^T(u_H^{(n)}))$ afterwards.

The remaining strategy is straightforward: we assemble $G(\alpha^{(n)})$ and solve the linear system $D_\alpha G(\alpha^{(n)})\triangle\alpha^{(n)} = -G(\alpha^{(n)})$. The subsequent HM Newton algorithm is a detailed summary of the whole procedure above.

Let $u_H^{(0)} \in V_H(\Omega)$ denote a suitable initial value for the HM Newton iterations. From now on, we use the following discretization $A_h^\varepsilon : \bigcup_{T\in\mathscr{T}_H} Y_{T,\delta} \times \mathbb{R}^N \to \mathbb{R}^N$ of the monotone elliptic diffusion operator A^ε. For $T \in \mathscr{T}_H,\ K \in \mathscr{T}_h$ and $\xi \in \mathbb{R}^N$ we define:

$$A_h^\varepsilon(\cdot,\xi)_{|x_T^\delta(K)} := A^\varepsilon(x_T^\delta(y_K),\xi) \text{ and}$$
$$(D_\xi A^\varepsilon)_h(\cdot,\xi)_{|x_T^\delta(K)} := (D_\xi A^\varepsilon)(x_T^\delta(y_K),\xi),$$

where $D_\xi A^\varepsilon$ denotes Jacobian matrix of A^ε with respect to the second variable. Now, for a given tolerance $TOL > 0$, we present our algorithm for solving the nonlinear elliptic HMM macro problem (6). It is in Petrov Galerkin formulation, which means that we do not reconstruct the test functions to save computational demand.

Algorithm [HM Newton Method]
Let $u_H^{(n)} \in V_H(\Omega)$ name the solution of the last HM Newton iteration step.

1. For $T \in \mathscr{T}_H$, determine the local corrector $Q_h^T(u_H^{(n)}) \in W_h(Y)$ defined by
$$\int_Y A_h^\varepsilon\left(x_T + \delta y, \nabla_x u_H^{(n)}(x_T) + \nabla_y Q_h^T(u_H^{(n)})(y)\right) \cdot \nabla_y \phi_h(y)\, dy = 0,$$
for all $\phi_h \in W_h(Y)$. Determine $Q_h^T(u_H^{(n)})$ for any $T \in \mathscr{T}_H$.
2. For any $T \in \mathscr{T}_H$ and for any macroscopic base function Φ_i $(1 \le i \le M)$ with $(\text{supp}\, \Phi_i) \cap T \neq \emptyset$, determine $D_{Q_h^T}(\Phi_i, u_H^{(n)}) \in W_h(Y)$ defined by:
$$\int_Y (D_\xi A^\varepsilon)_h\left(x_T + \delta y, \nabla_x u_H^{(n)}(x_T) + \nabla_y Q_h^T(u_H^{(n)})(x_T)\right)$$
$$\left(\nabla_x \Phi_i(x_T) + \nabla_y D_{Q_h^T}(\Phi_i, u_H^{(n)})(y)\right) \cdot \nabla_y \phi_h(y)\, dy = 0,$$
for all $\phi_h \in W_h(Y)$.
3. Define the entries of the HMM stiffness matrix $M^{(n)}$ by:
$$M_{ij}^{(n)} := \sum_{T\in\mathscr{T}_H} |T| \fint_{\frac{\varepsilon}{\delta}Y} \Big((D_\xi A^\varepsilon)_h(x_T + \delta y, \nabla_x u_H^{(n)}(x_T) + \nabla_y Q_h^T(u_H^{(n)})(y))$$
$$(\nabla_x \Phi_j(x_T) + \nabla_y D_{Q_h^T}(\Phi_j, u_H^{(n)})(y))\Big) \cdot \nabla_x \Phi_i(x_T)\, dy$$
and the entries of the right hand side by:

On the Analysis of Porous Media Dynamics Using a DUNE-PANDAS Interface

Maik Schenke and Wolfgang Ehlers

Abstract. The increasing complexity of numerical models and the requirement for more computational power is covered by multi-processor and multi-core hardware architectures. In order to exploit the capabilities of these machines, parallel algorithms are necessary. Therefore, the following article will describe first attempts in parallelizing the finite-element code PANDAS through an interface to the parallel software framework DUNE. In this regard, this connection will overcome the computational limits of the sequential finite-element code PANDAS by introducing its material definitions to the DUNE framework, in particular, to the external DUNE module DUNE-PDELAB. As a first example, the propagation of a two-dimensional elastic shear-wave through a fluid-saturated soil will be analyzed in a non-parallel environment.

1 Introduction

Porous materials can be found in several branches of engineering. They appear, for instance, as foams in mechanical engineering, as soil and concrete in civil engineering, or as soft and hard biological tissues, such as cartilage or bone, in biomechanical engineering. An adequate description of the geometry together with an accurate material description of porous media models generally results in complex numerical schemes. To satisfy the increasing requirements for computational power, multiple processors, which again may consist of multiple cores, are united in a single machine. In order to fully exploit the capabilities of these machines, parallel algorithms are necessary, which accomplish multiple tasks simultaneously.

The following article describes an interface between the parallel software framework DUNE, in particular, the external DUNE module DUNE-PDELAB, and the

Maik Schenke · Wolfgang Ehlers
Universität Stuttgart, Pfaffenwaldring 7, 70569 Stuttgart
e-mail: {maik.schenke,wolfgang.ehlers}@mechbau.uni-stuttgart.de

sequential finite-element code PANDAS, which allows for a parallelization of the latter, thus expanding its application range to more complex problems.

The requirements for a finite-element solver when analyzing coupled solid-fluid problems related to porous media, were not satisfied by any commercial finite-element package on the market in the early nineties [9]. Thus, the finite-element package PANDAS[1] was developed. Since then, PANDAS has been used to study different coupled problems, such as multi-component and multiphasic materials and with electro-chemical couplings under finite deformations [11, 12, 13, 14, 18]. PANDAS provides several interfaces to external tools, such as CUBIT and Gmsh for pre-processing, and to Tecplot and PANPost [17] for post-processing. The latter one is especially designed for PANDAS, however, it is based on the more general Tecplot data format. PANDAS, although written in C, exhibits a modular and object-oriented design.

The following article is structured as follows. Section 2 will shortly review a numerical soil model, which has been taken from the literature [9, 15] and which has been implemented in PANDAS. Section 3 will describe the current state of the interface between DUNE-PDELAB and PANDAS. It will provide details on the subroutines, the work flow, and the data exchange at the intersection of DUNE-PDELAB and PANDAS. Finally, the article will be concluded by a first numerical example in section 4. Therein, the propagation of an elastic shear-wave within a two-dimensional domain, which is described according to section 2, is analyzed.

2 An Incompressible Fluid-Saturated Soil Model

2.1 Theory of Porous Media

An adequate description of a fluid-saturated soil is provided by the Theory of Porous Media (TPM) [2, 6, 10], which proceeds from the Theory of Mixtures (TM) and the concept of volume fractions [1, 2, 3, 10, 19] . Within the macroscopic TPM approach, the porous media aggregate is treated as immiscible mixture φ of superimposed and interacting constituents φ^α. The solid skeleton ($\alpha = S$) and the pore fluid ($\alpha = F$) are assumed to be homogeneously distributed over a representative volume element (RVE) $\mathrm{d}v$, where each constituent occupies the partial volume $\mathrm{d}v^\alpha$ and is represented by the volume fraction n^α:

$$n^\alpha = \frac{\mathrm{d}v^\alpha}{\mathrm{d}v}, \qquad \sum_\alpha n^\alpha = \frac{1}{\mathrm{d}v}\sum_\alpha \mathrm{d}v^\alpha = 1\,. \tag{1}$$

Following this, two density functions, the material (realistic or effective) density $\rho^{\alpha R}$ and the partial (global or bulk) density ρ^α can be defined:

$$\rho^{\alpha R} = \frac{\mathrm{d}m^\alpha}{\mathrm{d}v^\alpha}, \qquad \rho^\alpha = \frac{\mathrm{d}m^\alpha}{\mathrm{d}v} \qquad \longrightarrow \qquad \rho^\alpha = n^\alpha \rho^{\alpha R}. \tag{2}$$

[1] **P**orous media **A**daptive **N**on-linear finite-element solver based on **D**ifferential **A**lgebraic **S**ystems [16].

It is seen from (2) that material incompressibility ($\rho^{\alpha R} = \text{const.}$) does not necessarily lead to bulk incompressibility, since the bulk density ρ^α can still change through a changing volume fraction n^α.

The kinematic relations of porous materials are described as superimposed continua, assuming that a spatial point $\mathcal{P}$ of the current configuration, defined by $\mathbf{x}$ at time t, is simultaneously occupied by particles of each constituent, where each particle is moving according to its own independent motion function $\boldsymbol{\chi}_\alpha$ starting from its reference position at t_0 (cf. Figure 2.1) given by $\mathbf{X}_\alpha$

$$\mathbf{x} = \boldsymbol{\chi}_\alpha(\mathbf{X}_\alpha, t) . \tag{3}$$

As a consequence, each constituent has its own independent velocity and accelera-

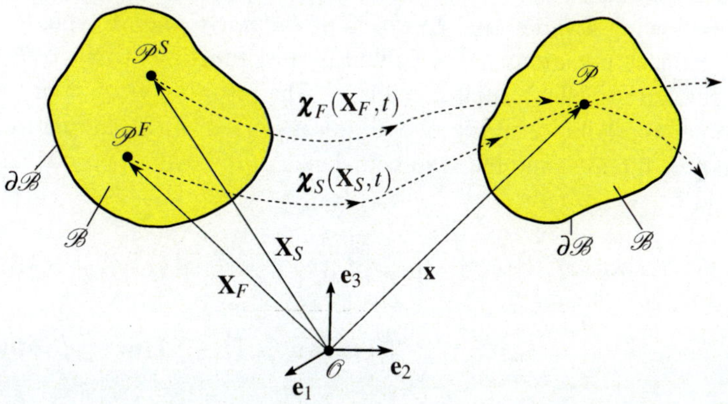

Fig. 1 Motion of a biphasic porous material consisting of a solid skeleton ($\alpha = S$) and pore fluid ($\alpha = F$).

tion field, viz.:

$$\overset{\prime}{\mathbf{x}}_\alpha = \frac{\mathrm{d}\boldsymbol{\chi}_\alpha(\mathbf{X}_\alpha,t)}{\mathrm{d}t} , \qquad \overset{\prime\prime}{\mathbf{x}}_\alpha = \frac{\mathrm{d}^2\boldsymbol{\chi}_\alpha(\mathbf{X}_\alpha,t)}{\mathrm{d}t^2} . \tag{4}$$

Therein, $(\cdot)'_\alpha$ denotes the material time derivative following the motion of φ^α:

$$(\cdot)'_\alpha := \frac{\mathrm{d}_\alpha(\cdot)}{\mathrm{d}t} = \frac{\partial(\cdot)}{\partial t} + \operatorname{grad}(\cdot)\cdot\mathbf{v}_\alpha \qquad \text{with} \qquad \operatorname{grad}(\cdot) = \partial(\cdot)/\partial\mathbf{x} . \tag{5}$$

Describing porous materials, it is convenient to express the solid motion in the *Lagrange*an or material setting by the displacement and velocity functions

$$\mathbf{u}_S = \mathbf{x}_S - \mathbf{X}_S , \qquad \mathbf{v}_S = (\mathbf{u}_S)'_S = \overset{\prime}{\mathbf{x}}_S , \tag{6}$$

and the fluid motion in the *Euler*ian or spatial setting by the seepage velocity $\mathbf{w}_F$ relative to the solid motion:

$$\mathbf{w}_F = \overset{\prime}{\mathbf{x}}_F - \overset{\prime}{\mathbf{x}}_S \qquad \text{with} \qquad \overset{\prime}{\mathbf{x}}_F = \mathbf{v}_F\,. \tag{7}$$

2.2 Balance Equations and Numerical Treatment

According to *Truesdell*'s metaphysical principles [19], the balance equations of classical continuum mechanics can be rewritten for an individual constituent, thereby including production terms to allow for an interaction between the solid skeleton and the pore fluid. Within the scope of this article, assuming iso-thermal conditions, the necessary balance relations are the momentum balance and the mass balance, where the latter reduces to a volume balance under the assumption of incompressible constituents.

Following a variational approach of *Bubnov-Galerkin* type, the strong forms of the balance equations are weighted by test functions ($\delta\mathbf{u}_S$, $\delta\mathbf{v}_F$ and δp) and are integrated over the spatial domain $\mathscr{B}$. The resulting weak forms, in particular, the momentum balance of the overall porous material (8), the momentum balance of the pore fluid (9) and the volume balance of the overall porous material (10), read, cf. [15]:

$$\begin{aligned} 0 = &\int_{\mathscr{B}} \delta\mathbf{u}_S \cdot \{\rho^S[(\mathbf{v}_S)'_S - \mathbf{b}] + \rho^F[(\mathbf{v}_F)'_S + (\operatorname{grad}\mathbf{v}_F)\mathbf{w}_F - \mathbf{b}]\}\,\mathrm{d}v + \\ &+ \int_{\mathscr{B}} \operatorname{grad}\delta\mathbf{u}_S \cdot (\mathbf{T}^S_E - p\,\mathbf{I})\,\mathrm{d}v - \int_{\partial\mathscr{B}_\mathbf{t}} \delta\mathbf{u}_S \cdot \bar{\mathbf{t}}\,\mathrm{d}a = 0\,, \end{aligned} \tag{8}$$

$$\begin{aligned} 0 = &\int_{\mathscr{B}} \delta\mathbf{v}_F \cdot [\rho^F(\mathbf{v}_F)'_S + (\operatorname{grad}\mathbf{v}_F)\mathbf{w}_F - \mathbf{b}]\mathrm{d}v + \int_{\partial\mathscr{B}_{\mathbf{t}^F}} \delta\mathbf{v}_F \cdot \bar{\mathbf{t}}^F\,\mathrm{d}a + \\ &+ \int_{\mathscr{B}} \delta\mathbf{v}_F \cdot [\frac{(n^F)^2\gamma^{FR}}{k^F}\mathbf{w}_F - p\operatorname{grad}n^F]\mathrm{d}v + \int_{\mathscr{B}} \operatorname{div}\delta\mathbf{v}_F(-n^F p)\mathrm{d}v = 0\,, \end{aligned} \tag{9}$$

$$\begin{aligned} 0 = &-\int_{\mathscr{B}} \operatorname{grad}\delta p \cdot \frac{k^F}{\gamma^{FR}}\{\rho^{FR}[\mathbf{b} - (\mathbf{v}_F)'_S - (\operatorname{grad}\mathbf{v}_F)\mathbf{w}_F] - \operatorname{grad}p\}\mathrm{d}v + \\ &+ \int_{\mathscr{B}} \delta p \operatorname{div}\mathbf{v}_S\,\mathrm{d}v + \int_{\partial\mathscr{B}_v} \delta p\,\bar{v}\,\mathrm{d}a = 0\,. \end{aligned} \tag{10}$$

Therein, $\bar{\mathbf{t}}$, $\bar{\mathbf{t}}^F$ and $\bar{v}$ denote the external load vector applied on the overall porous medium, the external load vector applied on the pore fluid, and the volume flux applied on the volume balance of the overall porous medium. $\bar{\mathbf{t}}$, $\bar{\mathbf{t}}^F$ and $\bar{v}$ are acting on the *Neumann* boundaries $\partial\mathscr{B}_\mathbf{t}$, $\partial\mathscr{B}_{\mathbf{t}^F}$ and $\partial\mathscr{B}_v$. Note that the unit surface normal is oriented outwards. Moreover, the solid displacement $\mathbf{u}_S = \overline{\mathbf{u}}_S$ on $\partial\mathscr{B}_{\mathbf{u}_S}$, the fluid velocity $\mathbf{v}_F = \overline{\mathbf{v}}_F$ on $\partial\mathscr{B}_{\mathbf{v}_F}$ and the pore pressure $p = \overline{p}$ on $\partial\mathscr{B}_p$ exactly fulfill the *Dirichlet* boundary conditions. Furthermore, $k^F > 0$ denotes the *Darcy* permeability and $\gamma^{FR} = \rho^{FR} g$ is the effective fluid weight with $g = |\mathbf{b}| = \text{const.}$ as the scalar gravitational acceleration.

In order to exploit the existing time-integration methods in DUNE-PDELAB, the order of (8) will be reduced by additionally satisfying $(6)_2$ in a strong sense.

To complete the model description, a constitutive relation for the solid extra stress $\mathbf{T}^S_E$ is necessary. Confined to the small strain regime

$$\boldsymbol{\varepsilon}_S \approx \frac{1}{2}\left(\operatorname{grad}\mathbf{u}_S + \operatorname{grad}^T\mathbf{u}_S\right), \tag{11}$$

the solid extra stress is sufficiently described by a *Hooke*an elasticity law

$$\mathbf{T}^S_E = 2\mu^S\,\boldsymbol{\varepsilon}_S + \lambda^S\,(\boldsymbol{\varepsilon}_S \cdot \mathbf{I})\,\mathbf{I}. \tag{12}$$

Therein, μ^S and λ^S denote the macroscopic *Lamé* constants of the porous solid skeleton.

Following this, a spatial discretization is carried out, thereby approximating the unknown fields $\mathbf{u}_S$ and $\mathbf{v}_S$ by quadratic shape functions, whereas linear shape functions are used for $\mathbf{v}_F$ and p in order to fulfill the *inf-sup* condition (*Ladyshenskaya–Babuška–Brezzi* (LBB) condition) [4] for the sake of stability. Finally, the semi-discrete initial-boundary-value problem can be summarized as

$$\boldsymbol{M}\boldsymbol{u}' + \boldsymbol{K}(\boldsymbol{u}) - \boldsymbol{f} = \mathbf{0}, \tag{13}$$

where $\boldsymbol{M}$ and $\boldsymbol{K}$ represent generalized mass and stiffness matrices, $\boldsymbol{f}$ the external load vector and $\boldsymbol{u}$ a vector containing the nodal degrees of freedom of the finite-element mesh. Note that the structure of (13) is similar to the structure of the ordinary differential equations presented in the DUNE-PDELAB manual [8].

3 A DUNE-PANDAS Interface

The following section describes the connection between the software framework DUNE, in particular, the external module DUNE-PDELAB, and the finite-element code PANDAS. When defining an interface between two software packages, one has to identify a common point, which is, within the current setting, the evaluation of the element contributions to the overall system of equations. In this regard, the computation of the material response is left up to PANDAS, from which DUNE-PDELAB gradually assembles the overall system of equations, applies the boundary conditions and finally solves the non-linear system of equations at every time step.

3.1 Interface Work Flow

The interface work flow incorporating both software packages is depicted in Figure 2. Following this, PANDAS is initialized at the beginning within `main()` of DUNE-PDELAB by `DunePandasInit()`. It builds up the data structure of PANDAS, obtains the material type and parameters according to a user-provided flag and assigns the test and trial function space to it.

The return value of `DunePandasInit()` is a pointer to a PANDAS mesh, which consists of a single element. Once the initialization of PANDAS is done, DUNE-PDELAB proceeds by gradually building up the grid function space, by defining the *Dirichlet* and the *Neumann* boundary conditions and by assigning a matrix and a PDE-solver back end. During an analysis, at each time step, DUNE-PDELAB assembles the complete system of equations from the element-wise contributions which are provided by the member functions `alpha_volume()` and `alpha_jacobian()` of the local operator and the time-local operator class. Within the current setting, the contributions to the overall system of equations are not evaluated by the member functions, instead, they are computed by PANDAS through `PhysResidual()`, contributing to the right-hand-side vector, and by `PhysTangent()`, contributing to the *Jacobi*an matrix. In order to proceed with the element-wise computations, the nodal coordinates of the DUNE element replace the nodal coordinates of the single-element PANDAS mesh. Note that the *Jacobian* matrix may be computed numerically or semi-analytically, depending on a user provided flag.

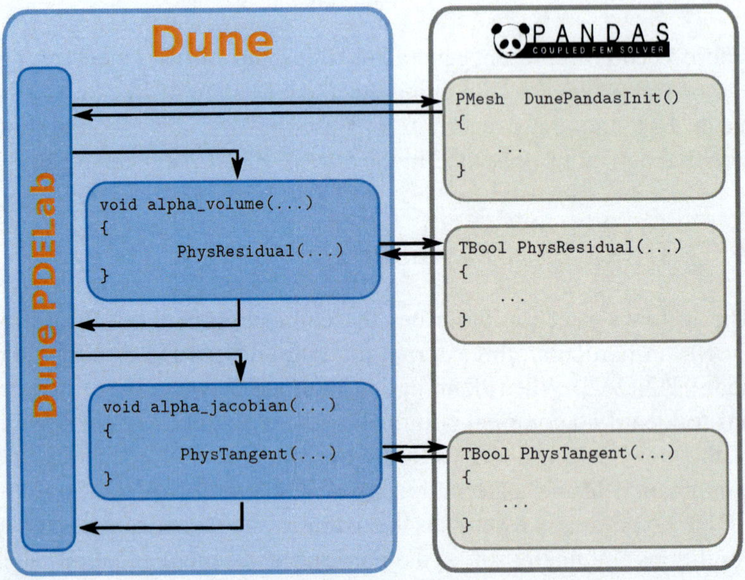

Fig. 2 DUNE work flow with connection to PANDAS.

Once the total system of equations is assembled, DUNE-PDELAB applies the *Dirichlet* and the *Neumann* boundary conditions and, finally, solves the non-linear system of equations by a suitable procedure (e. g. *Newton-Raphson* method) at every time step.

3.2 Data Transfer

As mentioned in the previous section, the common point in PANDAS and DUNE-PDELAB are the member functions of the local operator and the time-local operator class and the PANDAS subroutines `PhysResidual()` and `PhysTangent()`. The arguments of `alpha_volume()` and `PhysResidual()` are depicted in Figure 3, where a comparison shows several similarities.

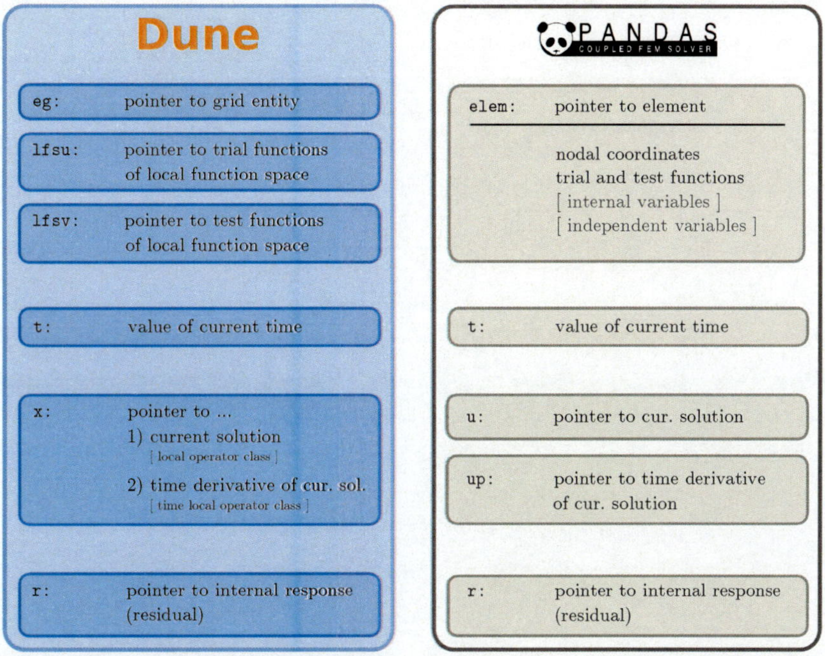

Fig. 3 Argument list of `alpha_volume()` (left) and `PhysResidual()` (right).

The argument `elem` of the PANDAS subroutine denotes a pointer to a C-structure representing a PANDAS element, which again contains references to the nodal coordinates of the trial- and test-function spaces, and to the storage spaces for internal variables (e. g. inelastic strains) and independent variables (e. g. stresses and seepage velocities), partially find their counterparts in the entity pointer `eg` and in the pointers to the trial- and test-function spaces (`lfsu` and `lfsv`). Following this, it is obvious that no storage spaces for internal and independent variables is available in DUNE-PDELAB. As a consequence, the usage of numerical models, where internal variables are essential for the computations at subsequent time steps,

An Implementation of Hybrid Discontinuous Galerkin Methods in DUNE

Christian Waluga and Herbert Egger

Abstract. We discuss the implementation of hybrid finite element methods in the Distributed and Unified Numerics Environment (DUNE) [5, 6]. Such hybrid methods require the approximation of the solution in the interior of the elements, as well as an approximation of the traces of the solution on the element interfaces, i.e., the skeleton. For illustration, we consider a hybrid version of the interior penalty discontinuous Galerkin method for an elliptic model problem. In order to realize the implementation in the DUNE framework, we present a generic extension of the C++ template library DUNE-PDELAB [4] for problems with additional polynomial approximations of spaces defined on the skeleton.

1 Introduction

Let Ω be a bounded polygonal or polyhedral domain in $\mathbb{R}^d$, $d = 2,3$ with boundary $\partial\Omega = \partial\Omega^D \cup \partial\Omega^N$ consisting of two regular components, a Dirichlet and Neumann part. We consider the Poisson problem with mixed boundary conditions: Find the weak solution $u \in H^1(\Omega)$ of

$$\begin{cases} -\Delta u = f, & \text{in } \Omega, \\ u = g, & \text{on } \partial\Omega^D, \\ \frac{\partial u}{\partial n} = h, & \text{on } \partial\Omega^N, \end{cases} \tag{1}$$

with given data $f \in L^2(\Omega)$, $g \in H^{1/2}(\partial\Omega^D)$, and $h \in L^2(\partial\Omega^N)$.

Christian Waluga
Aachen Institute for Advanced Study in Computational Engineering Science, RWTH Aachen University, Schinkelstraß e. 2, 52062 Aachen, Germany
e-mail: waluga@aices.rwth-aachen.de

Herbert Egger
Center for Mathematical Sciences, Technische Universität München, Boltzmannstraß e 3, 85748 Garching bei München, Germany
e-mail: herbert.egger@ma.tum.de

This problem serves as a model for various physical phenomena, e.g. electrostatics or diffusion processes. In heat conduction, for instance, u denotes the temperature distribution, f models internal heat sources, g is the prescribed temperature at the Dirichlet part, and h the heat flux over the Neumann part of the boundary.

2 A Hybrid Variational Principle

In the following, we propose a variational principle, which is satisfied by any regular solution of the Poisson problem (1), and which will be the basis for the hybrid finite element method studied in this paper.

2.1 *Preliminaries*

Let $\mathscr{T}_h = \{T\}$ be a partition of the domain Ω into a set of elements T, such as triangles or tetrahedral elements in two or three spaces dimensions, respectively. The intersections of codimension one of neighboring elements or of elements and the boundary, i.e. edges or faces for $d = 2$ or 3, respectively, are called facets. We denote by $\mathscr{E}_h^0 := \{E_{ij} = \partial T_i \cap \partial T_j : T_i, T_j \in \mathscr{T}_h,\ i > j\}$ the set of facets between neighboring elements, and by $\mathscr{E}_h^D := \{E_i : E_i = \partial T_i \cap \partial\Omega^D,\ T_i \in \mathscr{T}_h\}$ and $\mathscr{E}_h^N := \{E_i : E_i = \partial T_i \cap \partial\Omega^N,\ T_i \in \mathscr{T}_h\}$, the set of facets on the Dirichlet and Neumann boundary, respectively. The collection of all facets is denoted by $\mathscr{E}_h := \mathscr{E}_h^0 \cup \mathscr{E}_h^D \cup \mathscr{E}_h^N$, and the union $\mathscr{E} := \bigcup_{E \in \mathscr{E}_h} E$ of all element interfaces is referred to as the *skeleton*.

2.2 *A Weak Formulation*

For $s \geq 0$, let us introduce the broken Sobolev spaces

$$H^s(\mathscr{T}_h) := \{v \in L^2(\Omega) : v_h|_T \in H^s(T) \quad \forall T \in \mathscr{T}_h\} \tag{2}$$

defined on a partition $\mathscr{T}_h$ of the domain Ω, and set $V := H^2(\mathscr{T}_h)$. For $g \in L^2(\partial\Omega^D)$, we denote by

$$\widehat{V}^g := \{\hat{v} \in L^2(\mathscr{E}_h) : \hat{v}_h = g \quad \text{on } \partial\Omega^D\} \tag{3}$$

the set of functions defined over the skeleton with prescribed values on the Dirichlet boundary. As a weak form of the Poisson problem (1), we then consider the following *variational principle*: Find $(u, \hat{u}) \in V \times \widehat{V}^g$ such that

$$a_h(u, \hat{u}, v, \hat{v}) = f_h(v, \hat{v}), \qquad \forall (v, \hat{v}) \in V \times \widehat{V}^0, \tag{4}$$

with bilinear and linear forms defined by

$$
\begin{aligned}
a_h(u,\hat{u},v,\hat{v}) := &\sum_{T\in\mathscr{T}_h}\int_T \nabla u\cdot\nabla v\,dx - \sum_{T\in\mathscr{T}_h}\int_{\partial T}\frac{\partial u}{\partial n}(v-\hat{v})\,ds \\
&- \sum_{T\in\mathscr{T}_h}\int_{\partial T}(u-\hat{u})\frac{\partial v}{\partial n}\,ds + \sum_{T\in\mathscr{T}_h}\tau\int_{\partial T}(u-\hat{u})(v-\hat{v})\,ds,
\end{aligned}
$$

and

$$
f_h(v,\hat{v}) := \sum_{T\in\mathscr{T}_h}\int_T f\,v\,dx + \sum_{E\in\mathscr{E}_h^N}\int_E h\,\hat{v}\,ds.
$$

Here, n denotes the outward pointing normal vector on ∂T and τ is a piecewise constant positive stabilization parameter, which will be chosen later.

Proposition 1. *Let $u\in H^1(\Omega)\cap H^2(\mathscr{T}_h)$ denote a weak solution of the Poisson problem* (1)*, and define $\hat{u}=u|_{\mathscr{E}}$. Then $(u,\hat{u})$ also solves the variational problem* (4).

Proof. The result follows directly from the definition of the bilinear and linear forms, integration-by-parts, and noting that the flux $\frac{\partial u}{\partial n}$ is continuous across element interfaces, since we assumed $f\in L^2(\Omega)$.

Remark 1. The extra regularity of the solution u is only required, to ensure that the normal derivative $\frac{\partial u}{\partial n}$ is well defined on single edges. This requirement can be relaxed in several ways, e.g., $u\in H^{3/2+\varepsilon}(\mathscr{T}_h)$ is sufficient to obtain $\frac{\partial u}{\partial n}\in L^2(\mathscr{E}_h)$. Alternatively, one may define $V=H^1(\mathscr{T}_h)$ and $\widehat{V}^g=\{\hat{v}\in H^{1/2}(\mathscr{E}):\hat{v}=g \quad \text{on } \partial\Omega^D\}$, in which case any solution of the Poisson problem satisfies the variational principle. If the scalar product of $H^{1/2}(\mathscr{E})$ is utilized for penalization instead of the weighted L^2-term, then (4) completely characterizes weak solutions of (1), i.e., any solution of the variational formulation is also a solution of the Poisson problem [10].

3 Discretization by Finite Elements

For ease of presentation, we assume in the following, that $\mathscr{T}_h$ is a shape-regular, quasi-uniform simplicial mesh. Let us denote by $\mathscr{P}^p(T)$ and $\mathscr{P}^p(E)$ the space of polynomials of maximal order p on elements $T\in\mathscr{T}_h$ and facets $E\in\mathscr{E}_h$, respectively. For approximation of the variational principle (4), we utilize the following finite element spaces

$$
\begin{aligned}
V_h &:= \{v_h\in L^2(\Omega): v_h|_T\in\mathscr{P}^p(T) \quad \forall T\in\mathscr{T}_h\}, \qquad \text{and} \\
\widehat{V}_h &:= \{\hat{v}_h\in L^2(\mathscr{E}): \hat{v}_h|_E\in\mathscr{P}^p(E) \quad \forall E\in\mathscr{E}_h\}.
\end{aligned}
$$

In order to incorporate the Dirichlet boundary conditions, we further define the space

$$\widehat{V}_h^g := \{v \in \widehat{V}_h : v|_E = \pi_E g, \quad E \in \mathscr{E}_h^D\}, \tag{5}$$

where π_E is the orthogonal projection onto $\mathscr{P}^p(E)$. The space of functions, defined on the skeleton, which vanish on $\partial\Omega^D$ is denoted accordingly by $\widehat{V}_h^0$. Figure 1 displays the association of the degrees of freedom with elements and intersections for a linear approximation on a triangular mesh.

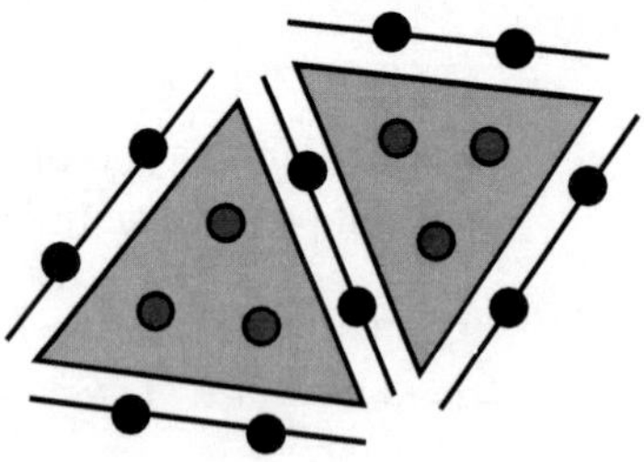

Fig. 1 Degrees of freedom associated with a hybridized discontinuous Galerkin approximation of first order. Black bullets indicate the globally coupled hybrid degrees of freedom and gray bullets denote (inner) element degrees of freedom.

3.1 *The Discrete Method*

As approximation for the Poisson problem, we then consider the following finite element discretization of the variational principle (4).

Problem 1 (Hybrid discontinuous Galerkin method). Given $f \in L^2(\Omega)$, $g \in H^{1/2}(\partial\Omega^D)$, and $h \in L^2(\partial\Omega^N)$, find $(u_h, \hat{u}_h) \in V_h \times \widehat{V}_h^g$ such that

$$a_h(u_h, \hat{u}_h, v_h, \hat{v}_h) = f_h(v_h, \hat{v}_h) \qquad \text{for all } v_h \in V_h \text{ and } \hat{v}_h \in \widehat{V}_h^0. \tag{6}$$

To ensure well-definedness of the method, we require that the stabilization parameter τ is chosen as $\tau \geq \alpha \frac{p^2}{h}$ with α depending on the shape regularity of the mesh.

Remark 2. Note that the Dirichlet boundary conditions are explicitly built into the hybrid solutions space $\widehat{V}_h^g$, and no conditions are imposed on the functions in V_h.

Remark 3. Problem 1 is a hybrid version of the symmetric interior penalty method [1]; non-symmetric versions can be defined accordingly. The idea of hybridization goes back at least to [2], where it was used to facilitate the implementation of mixed methods. The approach has been generalized to a large class of methods using discontinuous approximation spaces in [3]. The hybridization of discontinuous Galerkin methods has been investigated as a general framework in [8]; see also [10] for a-priori error estimates.

3.2 A-Priori Error Analysis

The analysis of the discrete problem is carried out with respect to the following mesh-dependent energy norm

$$\|(v,\hat{v})\|_{1,h}^2 := \sum_{T\in\mathscr{T}_h} \left(\|\nabla v\|_T^2 + \|\tau^{1/2}(v-\hat{v})\|_{\partial T}^2 \right).$$

The boundedness of a_h and f_h in the discrete norm follows immediately by the Cauchy-Schwarz inequality, a discrete Friedrichs-inequality, and discrete trace inequalities. The latter also imply the coercivity of the bilinear form a_h; for details, see [10].

Proposition 2. *Let α be sufficiently large. Then for all $(v_h,\hat{v}_h) \in V_h \times \widehat{V}_h^0$, there holds*

$$a_h(v_h,\hat{v}_h;v_h,\hat{v}_h) \geq \tfrac{1}{2}\|(v_h,\hat{v}_h)\|_{1,h}^2.$$

Based on Galerkin orthogonality, the error in the finite element solution can now be bounded by the best-approximation error, which yields the following estimates.

Theorem 1 (A-priori estimates). *Let $u \in H^{s+1}(\mathscr{T}_h)$ denote the solution of the Poisson problem* (1). *Then*

$$\|(u-u_h,u-\hat{u}_h)\|_{1,h} \leq C_s \frac{h^\mu}{p^{\mu-1/2}} \|u\|_{s+1,h},$$

with $\mu = \min\{p,s\}$ and constant C_s independent of h, p, and u.

Remark 4. Using the standard duality argument, one can also obtain order optimal rates with respect to the L^2-norm. For a-posteriori estimates derived in the spirit of [15, 14] for mixed and discontinuous Galerkin methods, see e.g. [12].

3.3 Additional Aspects

Let us make some remarks concerning generalizations and some implementational aspects of hybrid discontinuous Galerkin methods.

Remark 5 (Generalizations). For ease of presentation, we considered only the case of quasi-uniform simplicial meshes above. The a-priori analysis can easily be generalized to shape-regular meshes, varying polynomial degree, various element types, and hanging nodes; see e.g., [12]. The hybrid DG framework can even be used for mortaring [8, 10, 13].

Remark 6 (Choice of the stabilization parameter). While the stability of the hybrid method holds with a universal constant $1/2$, the constant C_s in the error estimate, and the conditioning of the resulting linear system is negatively affected, if the stabilization parameter α is chosen very large. Sharp explicit bounds in the discrete trace inequalities [20], however, allow to choose α in an optimal way.

Remark 7 (Static condensation). The discrete variational problem (6) has the form

$$\sum_{T\in\mathcal{T}_h} A_T(u_T,\hat{u}_T) = \sum_{T\in\mathcal{T}_h} f_T,$$

i.e., in contrast to standard discontinuous Galerkin methods, the assembling can be performed element-wise. Since the degrees of freedom associated with the primal space V_h do not couple over element interfaces, the primal variables u_h can be completely eliminated during assembly by *static condensation*. The global system then only involves the variables $\hat{u}_h$. In the simplest case, one can obtain a global system with the same sparsity pattern as a non-conforming P^1 discretization [11, 10].

4 Implementation in the DUNE Framework

In this section, we discuss the implementation[1] of the hybrid DG method, introduced in the previous section, in DUNE [5, 6]. An important ingredient here is the definition of degrees of freedom associated with the interface mesh $\mathcal{E}_h$. We therefore discuss the construction of the approximation space $\widehat{V}_h$ for the hybrid variable in some detail, and present a technique for incorporating the Dirichlet boundary conditions.

In the following, let `R` denote a real value type and `GV` a DUNE grid-view type. The constant `p` represents the polynomial degree p, which is here specified at compile-time. For illustration, we present short snippets of pseudocode; for ease of notation, we assume that the namespaces `Dune` and `Dune::PDELab` are used.

4.1 *Construction of the Discrete Discontinuous Galerkin Space V_h*

For the definition of the discontinuous Galerkin finite element space V_h, we make use of DUNE-PDELAB's [4] monomial finite element maps. This specialization of a finite element map defines degrees of freedom in the interior of each element $T \in \mathcal{T}_h$. The monomial shape functions of maximal order p are defined in a generic way on a d-dimensional reference element. An instantiation of V_h would look as follows:

```
typedef MonomLocalFiniteElementMap<R, R, d, p> VMap;
typedef GridFunctionSpace<GV, VMap, NoConstraints,
                              Backend> VSpace;

// create the FE map and space for V_h
VMap vmap(GeometryType::simplex);
VSpace vgfs(gv, vmap);
```

Note, that we do not constrain any of the degrees of freedom in V_h.

[1] The implementation presented here is based on the DUNE-module DUNE-HDG: `http://users.dune-project.org/projects/dunehdg`

4.2 Construction of the Discrete Hybrid Interface Space $\widehat{V}_h$

For the implementation of the hybrid finite element space $\widehat{V}_h$, we have to define degrees of freedom on the interface mesh $\mathscr{E}_h$. In order to do this, we implemented a new monomial finite element map, which provides monomial finite elements of co-dimension 1, whose degrees of freedom are associated with intersections.

The implementation, of this new finite element map is available in the class template `FacetMonomLocalFiniteElementMap` in DUNE-HDG. It is generic and allows all available element-geometry-types and an arbitrary order of approximation. Therefore, the spatial dimension d and the polynomial degree p have to be provided as template parameters for the finite element map. An instantiation of $\widehat{V}_h$ would look as follows:

```
typedef IntersectionIndexSet<GV> IIS;
typedef FacetMonomLocalFiniteElementMap<IIS, R, R, d, p> HVMap;
typedef GridFunctionSpace<GV, HVMap, FacetConstraints,
  Backend, GridFunctionStaticSize<IIS> > HVSpace;

IIS iis(gv); // the intersection index set

// create the FE map and space for hV_h
HVMap hvmap(iis, GeometryType::simplex);
HVSpace hvgfs(gv, hvmap, iis);
```

Let us remark that the grid function space is not created in the usual way. We have to provide an intersection index set, which is implemented in DUNE-PDELAB, as a parameter to the `GridFunctionSpace` object. The specialized implementation of the grid function space is then able to define degrees of freedom associated with facets in our mesh. Also, since we will later constrain the boundary degrees of freedom in $\widehat{V}_h$, we add the newly developed `FacetConstraints` to the template parameters.

4.3 Construction of the Composite Space $V_h \times \widehat{V}_h$

The generic assembly procedures in DUNE-PDELAB expect a single grid function space as an argument. We thus combine the two previously defined grid function spaces in a standard composite grid function space as follows:

```
typedef CompositeGridFunctionSpace<GFSMapper,
                                           VSpace, HVSpace> GFS;
GFS gfs(vgfs, hvgfs);   // create the FE space V_h x hV_h
```

4.4 Boundary Conditions

While the Neumann boundary conditions are imposed in a weak sense, the Dirichlet boundary conditions are built into the finite element space $\widehat{V}_h^g$. We do this by constraining the corresponding degrees of freedom associated with the boundary facets by using the `FacetConstraints` as a parameter in the definition of $\widehat{V}_h$, cf. section 4.2.

In order to know, which boundary degrees of freedom need to be constrained, we need a function that evaluates the type of boundary condition at a point $x \in \partial\Omega$. As we have a composite space, we also need a composite grid function here.

```
typedef DummyBoundaryGridFunction<GV>      BCV;
typedef DirichletBoundaryGridFunction<GV> BChV;
typedef CompositeGridFunction<BCV, BChV>   BC;

BCV  bcv(gv);    // dummy constraints for V_h
BChV bchv(gv);   // Dirichlet constraints for hV_h

BC   bc(bcv, bchv); // composite for V_h x hV_h
```

For the first component, which remains unconstrained, we provide a dummy boundary grid function that does nothing. The second component, i.e. the hybrid part of our function space, has Dirichlet boundary conditions on the whole of $\partial\Omega$ in the example above. The constraints map is the created in the usual way:

```
typedef GFS::ConstraintsContainer<R>::Type C;
C constr;
constraints(bc, gfs, constr);
```

The actual values of the coefficients to be constrained can be determined using the helper function `interpolate_trace`, which can be used to project g to the discrete space $\widehat{V}_h$. This function mimics the behavior of DUNE-PDELAB's function `interpolate`, but works on the traces in $\mathscr{E}_h$ of elements. We apply this function only to the second sub-space of our composite grid function space:

```
typedef GFS::VectorContainer<R>::Type V;
V x(gfs, 0.0); // initialize vector of unknowns
G g(gv);       // function type: u=g on boundary

// select first sub-space from composite
typedef GridFunctionSubSpace<GFS, 1> HVSubSpace;
HVSubSpace hvsub(gfs);

// determine the constrained coefficients
interpolate_trace(g, hvsub, x, false);
```

In order to avoid unnecessary interpolations here, the last parameter `doSkeleton` of `interpolate_trace` can be set to **`false`**, which means that only boundary coefficients are determined.

4.5 The Residual Operator

The implementation of a local operator in DUNE-PDELAB is based on a weighted residual formulation: Find $(u_h, \hat{u}_h) \in V_h \times \widehat{V}_h^g$, such that

$$\mathscr{R}_h(u_h, \hat{u}_h, v_h, \hat{v}_h) = 0, \qquad \forall (v_h, \hat{v}_h) \in V_h \times \widehat{V}_h^0,$$

where $\mathscr{R}_h(u_h, \hat{u}_h, v_h, \hat{v}_h) := a_h(u_h, \hat{u}_h, v_h, \hat{v}_h) - f_h(v_h, \hat{v}_h)$. This can be split up into localized contributions, that can be separately assembled on each volume element in $\mathscr{T}_h$, inner intersection in ${\mathscr{E}_h}^0$ and boundary intersection, in ${\mathscr{E}_h}^D$ and ${\mathscr{E}_h}^N$.

Remark 8. The current implementation of our local operators computes the contribution of the two neighboring elements T_i and T_j to the integrals over an intersection $E_{ij} \in \mathscr{E}_h^0$ at once. This is due to the fact that we need a uniquely defined orientation in our intersection. The current DUNE-Interface however cannot guarantee that an intersection E_{ij} has the same orientation as its counterpart E_{ji} in the neighboring element. This can be seen as an disadvantage of the current implementation, since we sacrifice the locality of our element matrix assembly. The possibility of an efficient implementation of the static condensation procedure (cf. Remark 7) is thus clearly limited.

The implementation of the local operator for the Poisson problem is based on the discrete variational problem (6), and can be found in the template classes `HDGPoisson` and `HDGPoissonJacobian` in the DUNE-HDG module. The functions τ, f and j are template parameters for the local operator and need to be specified for an actual problem. For the computation of τ, there are several default implementations, including an optimal version `OptimalStabilityParameter`, which computes τ with help of the exact expressions of α, which are known for simplex and hypercube elements in arbitrary dimensions (cf. Remark 6).

As an example, we show the creation of the grid operator space here:

```
typedef ConstantFunction<GV, R> CF;
typedef OptimalStabilityParameter<GV> Tau;

CF f(gv, 0.0);      // source term: f=0
CF j(gv, 0.0);      // Neumann: j=0 here
Tau tau(p, alpha); // stability parameter

typedef HDGPoissonJacobian<CF, BC, CF, Tau> LocalOp;
typedef GridOperatorSpace<GFS, GFS, LocalOp,
                          C, C, Backend> GOS;

// create local operator and grid operator space
LocalOp lop(f, bc, j, t);
GOS gos(gfs, constr, gfs, constr, lop);
```

4.6 Additional Remarks

Our implementation naturally supports non-matching meshes, since we require that we have a sufficient number of degrees of freedom on each intersection. Using a suitable local error estimator, we can then employ the standard mesh adaption routines implemented in DUNE to obtain a generic h-adaptive finite element code; cf. e.g. [12].

The system is then assembled in the usual way, which is described in [4]. We solve the problem with a linear solver provided in the iterative solver template library DUNE-ISTL [7].

5 Numerical Results

In this section we present some numerical results that were obtained using the DUNE-HDG implementation on a triangular ALUGRID [9, 19].

We consider a stationary irrotational and incompressible flow velocity field $u(x,y)$ around a sphere with a radius $R = 1/4$ in two dimensions. Since $\nabla \times u = 0$ by assumption, there exists a velocity potential ψ such that $\nabla \psi = (u_2, -u_1)$. In the case of an incompressible flow, the velocity potential ψ then satisfies Laplace's equation: $-\Delta \psi = 0$. We solve for the velocity potential on $\Omega := (-1,1) \times (-\frac{1}{2}, \frac{1}{2})$ using the hybrid discontinuous Galerkin formulation presented in the previous sections. The boundary conditions are chosen in agreement with the exact solution

$$\psi(x,y) = y\frac{x^2 + y^2 - R^2}{x^2 + y^2}.$$

Figure 2 shows a solution plot for ψ on a triangular mesh. Note that the streamlines of the velocity field $u(x,y)$ coincide with the isolines of its potential function $\psi(x,y)$. In Table 1, we display the numerically observed convergence rates for $p = 2$, which are in good agreement with the theoretical predictions of Theorem 1.

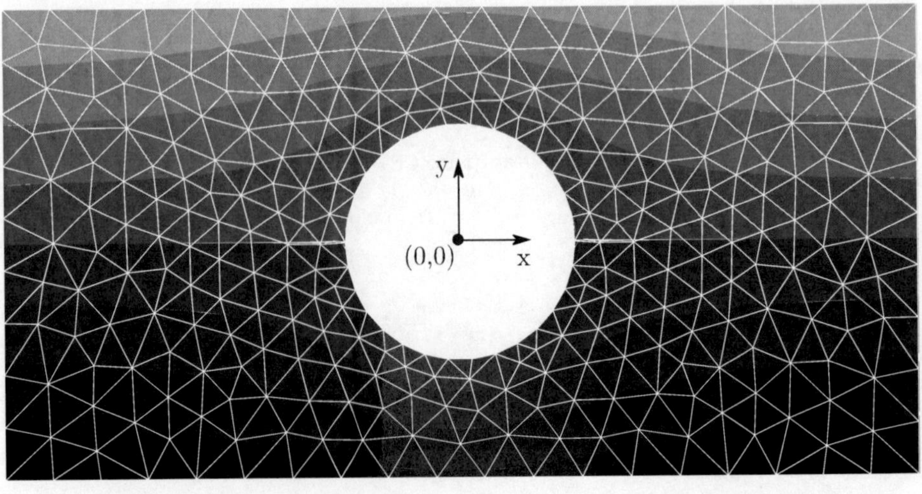

Fig. 2 Triangular mesh (level 0) and plot of ψ for the potential flow around a sphere.

Table 1 Errors of the numerical solution of the potential flow problem for a second order finite element approximation, and a sequence of uniformly refined meshes.

level	L^2	rate	energy	rate
0	$3.22565 \cdot 10^{-5}$	–	$4.64175 \cdot 10^{-3}$	–
1	$4.04213 \cdot 10^{-6}$	3.00	$1.19992 \cdot 10^{-3}$	1.95
2	$5.02869 \cdot 10^{-7}$	3.01	$3.03856 \cdot 10^{-4}$	1.98

6 Conclusion

We investigated the implementation of a hybrid discontinuous Galerkin method of a scalar elliptic model problem. Hybrid methods are also available for various other problems, including incompressible flow [17, 13, 12], convection-diffusion problems [11, 18], and the Helmholtz equations [16]. The framework proposed in this paper should allow an easy implementation also for these problems.

The proposed implementation naturally supports nonconforming adaptive meshes, and allows the extension to a locally varying polynomial degree (p-adaption). These cases can also be covered by the presented analysis; for details, see e.g. [12].

One of the basic features of the hybrid discontinuous Galerkin method, is the element-wise assembly process, which provides the possibility for static condensation on the element level. We were not able to implement the assembling as a single loop over elements in the current version of DUNE. We intend however, to further investigate this issue. The static condensation on the element level, and the resulting assembly of the reduced global (Schur complement) system will probably require the implementation of a specialized grid operator space in DUNE-PDELAB.

Acknowledgements. The first author would like to thank Steffen Muething (University of Stuttgart) for fruitful discussions, which lead to several improvements of the implementation, also Andreas Dedner (Warwick Mathematics Institute) and the users of the DUNE mailing list, for helpful comments and hints. Both authors gratefully acknowledge financial support by the DFG through grant GSC 111 and from FWF through grant SFB/F032.

References

1. Arnold, D.N.: An interior penalty finite element method with discontinuous elements. SIAM J. Numer. Anal. 19, 742–760 (1982)
2. Arnold, D.N., Brezzi, F.: Mixed and nonconforming finite element methods: Implementation, postprocessing and error estimates. Math. Model. Numer. Anal. 19, 7–32 (1985)
3. Baiocchi, C., Brezzi, F., Marini, L.D.: Stabilization of Galerkin Methods and Applications to Domain Decomposition. In: Bensoussan, A., Verjus, J.-P. (eds.) INRIA 1992. LNCS, vol. 653, pp. 343–355. Springer, Heidelberg (1992)
4. Bastian, P., Heimann, F., Marnach, S.: Generic implementation of finite element methods in the Distributed and Unified Numerics Environment (DUNE). Kybernetika 46(2), 294–315 (2010)

5. Bastian, P., Blatt, M., Dedner, A., Engwer, C., Klöfkorn, R., Ohlberger, M., Sander, O.: A Generic Grid Interface for Parallel and Adaptive Scientific Computing. Part I: Abstract Framework. Computing 82(2-3), 103–119 (2008)
6. Bastian, P., Blatt, M., Dedner, A., Engwer, C., Klöfkorn, R., Kornhuber, R., Ohlberger, M., Sander, O.: A Generic Grid Interface for Parallel and Adaptive Scientific Computing. Part II: Implementation and Tests in DUNE. Computing 82(2-3), 121–138 (2008)
7. Blatt, M., Bastian, P.: The Iterative Solver Template Library. In: Kågström, B., Elmroth, E., Dongarra, J., Waśniewski, J. (eds.) PARA 2006. LNCS, vol. 4699, pp. 666–675. Springer, Heidelberg (2007)
8. Cockburn, B., Gopalakrishnan, J., Lazarov, R.: Unified hybridization of discontinuous Galerkin, mixed and continuous Galerkin methods for second order elliptic problems. SIAM J. Numer. Anal. 47(2), 1319–1365 (2009)
9. Dedner, A., Rohde, C., Schupp, B., Wesenberg, M.: A parallel, load balanced MHD code on locally adapted, unstructured grids in 3D. In: Computing and Visualization in Science. Lecture Notes in Scientific Computing, vol. 7, pp. 79–96 (2004)
10. Egger, H.: A class of hybrid mortar finite element methods for interface problems with non-matching meshes. Preprint AICES-2009-2 (January 2009)
11. Egger, H., Schöberl, J.: A mixed-hybrid-discontinuous Galerkin finite element method for convection-diffusion problems. IMA J. Numer. Anal. (January 2009)
12. Egger, H., Waluga, C.: Hp-Analysis of a Hybrid DG Method for Stokes Flow. Submitted, Preprint AICES-2011/04-2 (April 2011)
13. Egger, H., Waluga, C.: A hybrid mortar method for incompressible flow. To appear in International Journal of Numerical Analysis and Modeling (2012)
14. Houston, P., Schötzau, D., Wihler, T.P., Schwab, C.: Energy norm a posteriori error estimation of hp-adaptive discontinuous Galerkin methods for elliptic problems. Mathematical Models and Methods in Applied Sciences 17(1), 33–62 (2007)
15. Kim, K.Y.: A posteriori error analysis for locally conservative mixed methods. Mathematics of Computation 76(257), 43 (2007)
16. Monk, P., Schöberl, J., Sinwel, A.: Hybridizing Raviart-Thomas elements for the Helmholtz equation. Electromagnetics 30(1), 149–176 (2010)
17. Nguyen, N.C., Peraire, J., Cockburn, B.: A hybridizable discontinuous Galerkin method for Stokes flow. Computer Methods in Applied Mechanics and Engineering 199(9-12), 582–597 (2010)
18. Nguyen, N.C., Peraire, J., Cockburn, B.: An implicit high-order hybridizable discontinuous Galerkin method for linear convection-diffusion equations. Journal of Computational Physics 228(9), 3232–3254 (2009)
19. Schupp, B.: Entwicklung eines effizienten Verfahrens zur Simulation kompressibler Strömungen in 3D auf Parallelrechnern. Dissertation, Mathematische Fakultät, Universität Freiburg (1999)
20. Warburton, T., Hesthaven, J.S.: On the constants in hp-finite element trace inverse inequalities. Computer Methods in Applied Mechanics and Engineering 192(25), 2765–2773 (2003)

Author Index

Subject Index